THE ORGANIZATION OF KNOWLEDGE: CAUGHT BETWEEN GLOBAL STRUCTURES AND LOCAL MEANING

STUDIES IN INFORMATION (PREVIOUSLY LIBRARY AND INFORMATION SCIENCE)

Series Editor: **Jens-Erik Mai**

Recent Volumes (Library and Information Science)

Dania Bilal and Jamshid Beheshti
New Directions in Children's and Adolescents' Information Behavior Research

Christine Bruce, Kate Davis, Hilary Hughes, Helen Partridge and Ian Stoodley
Information Experience: Approaches to Theory and Practice

Mark Hepworth and Geoff Walton
Developing People's Information Capabilities: Fostering Information Literacy in Educational, Workplace and Community Contexts

Jung-Ran Park and Lynne C. Howarth
New Directions in Information Organization

Amanda Spink and Jannica Heinström
Trends and Research: Europe

Gunilla Widén and Kim Holmberg
Social Information Research

Dirk Lewandowski
Web Search Engine Research

Donald Case
Looking for Information, Third Edition

Amanda Spink and Diljit Singh
Trends and Research: Asia-Oceania

Amanda Spink and Jannica Heinstrom
New Directions in Information Behaviour

Eileen G. Abels and Deborah P. Klein
Business Information: Needs and Strategies

Leo Egghe
Power Laws in the Information Production Process: Lotkaian Informetrics

Mike Thelwall
Link Analysis: An Information Science Approach

Matthew Locke Saxton and John V. Richardson
Understanding Reference Transactions: Transforming Art into a Science

Robert M. Hayes
Models for Library Management, Decision-Making, and Planning

Charles T. Meadow, Bert R. Boyce and Donald H. Kraft
Text Information Retrieval Systems, Second Edition

A. J. Meadows
Communicating Research

V. Frants, J. Shiparo, and V. Votskunskii
Automated Information Retrieval: Theory and Methods

STUDIES IN INFORMATION

THE ORGANIZATION OF KNOWLEDGE: CAUGHT BETWEEN GLOBAL STRUCTURES AND LOCAL MEANING

EDITED BY

JACK ANDERSEN
Royal School of Library and Information Science, University of Copenhagen, Copenhagen, Denmark

LAURA SKOUVIG
Royal School of Library and Information Science, University of Copenhagen, Copenhagen, Denmark

Series Editor: Jens-Erik Mai

emerald PUBLISHING

United Kingdom – North America – Japan
India – Malaysia – China

Emerald Publishing Limited
Howard House, Wagon Lane, Bingley BD16 1WA, UK

First edition 2017

British Library Cataloguing in Publication Data
A catalogue record for this book is available from the British Library

ISBN: 978-1-78714-532-0 (Print)
ISBN: 978-1-78714-531-3 (Online)
ISBN: 978-1-78714-964-9 (Epub)

ISSN: 2055-5377 (Series)

Printed and bound by CPI Group (UK) Ltd, Croydon, CR0 4YY

ISOQAR
REGISTERED
Certificate Number 1985
ISO 14001

ISOQAR certified Management System, awarded to Emerald for adherence to Environmental standard ISO 14001:2004.

INVESTOR IN PEOPLE

Contents

List of Contributors

Melissa Adler	School of Information Science, University of Kentucky, Lexington, KY, USA
Jack Andersen	Royal School of Library and Information Science, University of Copenhagen, Copenhagen, Denmark
José Augusto Chaves Guimarães	Graduate School of Information Science, UNESP – São Paulo State University, São Paulo, Brazil
Daniel Martínez-Ávila	Graduate School of Information Science, UNESP – São Paulo State University, São Paulo, Brazil
Fabio Assis Pinho	Department of Information Science, Federal University of Pernambuco, Recife, Brazil
Laura Skouvig	Royal School of Library and Information Science, University of Copenhagen, Copenhagen, Denmark

Introduction

This book is the result of a one-day conference held in Copenhagen in August 2015. The organizer of the conference was Professor Jens-Erik Mai, University of Copenhagen.

The theme of the conference was Global and Local Knowledge Organization and it was framed thus:

> Contemporary digital information society has globalized information structures and facilitated easier access to information across libraries, cultural institutions, and the Internet. While this has helped shaped global discourses, it has often done so at the expense of localized meaning and ethics. This conference sparks a conversation about the tension between the global information structures and grounding meaning and ethics in localized contexts.

In a world where the global and the local constantly intermingle, it was opportune to inquire into how knowledge organization research deals with the problems and challenges caused by the interaction of global and local information infrastructures. Beginning to look into what knowledge organization as activity and practice does in global and local information infrastructures is of course a challenging commitment. As such this book is also the result of an examination of what kinds of questions can be posed and addressed regarding issues of global and local information infrastructures. Ideas of universal bibliographical control and universal classification systems have been with us for centuries, from Conrad Gesner's *Bibliotheca Universalis* through Paul Otlet's UDC-system to Google's vision of organizing the world's information. These ideas and visions always face the same problem concerning the tensions and contradictions between devising universal systems and the local appropriation of these systems. Thus, we should not expect, nor work for, a perfect solution to this problem as human meaning construction, symbolic interactions, and various forms of ideologies will always be part of the picture. Therefore, the most challenging and pertinent task for knowledge organization research is to constantly and critically reflect upon the complexities of human meaning construction, symbolic interactions, and ideologies at stake. These take on different articulations and content depending on the kinds of questions we ask and from where in time and space we ask these questions. The texts in this volume are an attempt at doing this.

The activity and practice of knowledge organization has during the last 10 to 15 years moved out of the professional walls of libraries, archives, or the like institutions. That is of course not to say that the professional work carried out by these institutions has become irrelevant. It is, however, to recognize the fact that the activity and practice of knowledge organization is now something most of us do on a regular basis: we tag, structure, archive, or search structured digital collections because they are embedded in various forms of software and digital media. Thus, to organize knowledge is not restricted to professional domains. It is also a way of getting around in and making sense of digital information infrastructures. This opens new pathways as to what kinds of questions, conversations, and analytical frameworks to work with in knowledge organization research. For instance, studying the social and cultural implications of tagging, ordering, and archiving culture and cultural products has led some scholars from media, communication, and cultural studies to work with problems of knowledge organization (Beer, 2013; Gillespie, 2014; Hallinan & Striphas, 2016). These scholars work with knowledge organization exactly because the activity and practice has moved into the cultural sphere; that it is something people do in their everyday life in order to engage with digital information infrastructures. But also because knowledge organization in terms of algorithms has implications for how we as individuals are described, how cultural products are presented and circulated, and how public opinion is shaped by code, data, and algorithms — until recently features of software that did not impinge on social and public affairs. But now they do because of the omnipresence of software and digital media and the networked and platformed nature of communication.

However, this circumstance also requires a particular commitment by those who have been engaged in knowledge organization research in information studies. It is an invitation to step into the conversation and inform and shape modern ideas of tagging, ordering, or archiving in terms of what are the pitfalls and what we do already know about, for instance, the social, cultural, and ideological consequences of ordering and categorization from studies of library classification. There is a strong record of research to draw upon here (Bowker & Star, 1999; Mai, 2013; Olson, 2002; Wilson, 1968) and we should offer that line of thinking to the modern contexts. We should see this as an impetus for knowledge organization research to reach out and inform an agenda about what directions our digital information infrastructures can and should take.

Moreover, due to the omnipresence of classification and categorization in today's digital information infrastructures, questions about knowledge organization become questions with social, cultural, and political ramifications in broader spheres of activity than libraries. Whereas studies of library classification could point out how various people and subjects were described according to some hegemonic ideology (Berman, 1971; Olson, 2002), nowadays almost all our mundane activity is being classified and described because of our daily interactions with and reliance on software and digital media. That does not diminish the issue of ideology in classification. It becomes an issue that applies to all of us as citizens, private persons, or employees because our digital actions are archived, calculated, and classified by not only the state, as in Foucault's biopolitics (Foucault, 2008), but also by private companies, schools, universities, individuals, and various digital platforms. Accordingly, questions about knowledge organization become questions deeply intertwined with commerce, surveillance, identity, cultural taste, or political actions — all spheres that are much broader than library classification and with direct social, cultural, and political implications. This situation puts knowledge organization research near the center in studies of digital culture, provided that the field responds with adequate and socially relevant questions, analyses, and critiques.

The contributions in this volume are all written by scholars in information studies. From their respective points of departure, all the chapters are suggestive in terms of how to deal with modern problems of classification, categorization, and description. They all shape and critically analyze in both historical terms and contemporary settings how the organization of knowledge is a powerful means of understanding human meaning construction, symbolic interaction, and ideologies at stake in present-day digital information infrastructures.

The opening piece of the volume is written by Jack Andersen. He sets out to offer a twofold understanding of the organization of knowledge as both a form of communicative action in digital culture and as means to understand features of digital culture. With this, Andersen offers a new way to analytically approach studies of digital culture.

The following two chapters are historical in their orientation. Laura Skouvig suggests a Foucauldian genealogy in order to carry out a critical attitude toward our contemporary understandings of information. In a reading of different works on early modern information cultures, she discusses how different conceptualizations of information affect concrete uses of technologies of information.

A general assumption in Melissa Adler's chapter is that the way Library of Congress has classified people, nations, and territories must also be viewed as a history of how the United States has researched, used, or organized them into U.S. history. In her historical-genealogical study of the treatment of Māori History in the Library of Congress Classification (LCC), Adler examines how the United States is the universalized and assumed norm guiding the classification of subjects in LCC and how this can be a starting point for interrogating the intended and unintended consequences of universal classification schemes.

The next three chapters are written by the Spanish and Brazilian scholars Daniel Martínez-Ávila, Fabio Assis Pinho, and José Augusto Chaves Guimarães.

Martínez-Avila is looking at reader-interest classifications in order to consider the social consequences of such classifications and questions whether reader-interest classifications really consider the totality of users.

Fabio Assis Pinho's chapter is concerned with the problems at the intersection of knowledge representation and institutional memory. Pinho considers how a field of description of the informational content of photography allows its contextualization to contribute to the construction of institutional memory. Pinho's main argument is that there is a need to contextualize photographic documents in local systems in order to be able to construct a corresponding institutional memory.

In his chapter on slanted knowledge organization, José Augusto Chaves Guimarães argues that the organization of knowledge is a naturally slanted field. As such, Guimarães argues, this slanted nature of knowledge organization should be recognized as an ethical option in the theory and practice of knowledge organization.

Overall, this volume presents discussions that address the tensions and contradictions between global and local information infrastructures. In a digital and globalized world, human sense-making activities, for example, classification, are called upon and continuously contested. Thus, knowledge organization research must also continuously deal with these challenges and demonstrate how they can be dealt with practically, analytically, empirically, and theoretically. Fundamentally, to classify (or to tag, order, archive, sort) is always an articulation of the relationship between the classifier (whether human or algorithm) and the classified. This relationship is a political one as decisions have to be made, boundaries have to be drawn, and categories have to be established. From various approaches and perspectives, the chapters in this volume have centered on this relationship and have articulated their different takes on this issue. Consequently, we can make use of all the chapters to inform about why the problems of classification will always be with us, why the problems of classification are political, and why the problems of classification are socially and culturally embedded. The chapters can help to shape a contemporary conversation about where the ideas of classification come from and what directions they can possibly take.

Jack Andersen
Laura Skouvig
Editors

References

Beer, D. (2013). *Popular culture and new media. The politics of circulation.* New York, NY: Palgrave Macmillan.

Berman, S. (1971). *Prejudices and antipathies: A tract on the LC subject Heads concerning people.* Metuchen, NJ: Scarecrow.

Bowker, G., & Star, S. L. (1999). *Sorting things out. Classification and its consequences.* Cambridge, MA: MIT Press.

Foucault, M. (2008). *The birth of biopolitics: Lectures at the Collége de France, 1978-1979.* New York, NY: Palgrave Macmillan.

Gillespie, T. (2014). The relevance of algorithms. In T. Gillespie, P. J. Boczkowski, & K. A. Foot (Eds.), *Media technologies: Essays on communication, materiality, and society* (pp. 167–193). Cambridge, MA: MIT Press.

Hallinan, B., & Striphas, T. (2016). Recommended for you: The Netflix Prize and the production of algorithmic culture. *New Media & Society*, *18*(1), 117–137.

Mai, J. E. (2013). Ethics, values and morality in contemporary library classifications. *Knowledge Organization*, *40*(3), 242–253.

Olson, H. A. (2002). *The power to name: Locating the limits of subject representation in libraries.* Dordrecht: Kluwer.

Wilson, P. (1968). *Two kinds of power. An essay on bibliographical control.* Los Angeles, CA: University of California Press.

Chapter 1

Genre, Organized Knowledge, and Communicative Action in Digital Culture

Jack Andersen

Abstract

The purpose of the chapter is to argue for a twofold understanding of knowledge organization: the organization of knowledge as a form of communicative action in digital culture and the organization of knowledge as an analytical means to address features of digital culture.

The approach taken is an interpretative text-based form of argumentation.

The chapter suggests that by putting forward such a twofold understanding of knowledge organization, new directions are given as to how to situate and understand the activity and practice of the organization of knowledge in digital culture.

By offering the twofold understanding of the organization of knowledge, a tool of reflection is provided when users and the public at large try to make sense of, for example, data, archives, search engines, or algorithms.

The originality of the chapter is its demonstration of how to conceive of knowledge organization as a form of communicative action and as an analytical means for understanding issues in digital culture.

Keywords: The organization of knowledge; genre; digital culture; digital media; communicative action

---

The Organization of Knowledge: Caught Between Global Structures and Local Meaning
Studies in Information, Volume 12, 1–16

ISSN: 2055-5377/doi:10.1108/S2055-537720170000011007

Chapter Outline

1.1. Introduction

Any given form of culture can always be described in a variety of ways: a literate culture, an enlightenment culture, a culture of participation, a culture of experience, or as "a whole way of life" (Williams, 1989). Our current form of digital media culture, I suggest, can be described as a culture of ordering, structuring, and archiving as we witness an abundance of activities and practices that have to do with organizing, listing, archiving, ordering, or searching for items; that is, what is in information studies called the organization of knowledge. What used to be the professional practice of archivists and librarians, is now a way of navigating media culture and of producing culture, whether by humans or algorithms. The activities and practices of ordering, listing, archiving, categorizing, and searching are carried out by people in their everyday interactions in digital networks, suggesting that the organization of knowledge is a tool used to make sense of our daily routinized communicative interactions. Moreover, given the omnipresence of social, mobile, and networked media in our everyday life, it is too a tool employed to make sense of our lifeworld as it stretches from local, through regional, and to global digital contexts.

Moving on from these observations, we can begin to question how the organization of knowledge shapes everyday life because of its "inscription" in digital media. In digital culture, ordering and archiving is a way of interacting with digital media. It is "a whole way of life." It is a form of communicative action: One cannot experience digital media without also practicing knowledge organization: people construct queries, examine results, and access retrieved items as a matter of course when interacting with digital content. Accordingly, the creation of data to facilitate search and retrieval (tagging, labeling, or linking) is likewise a communicative action.

In this chapter, I will try to take a step back and think about how we can describe and understand ordering and archiving and its place and meaning in digital culture. In doing so, I will suggest a twofold understanding of the organization of knowledge. First, I will elaborate an understanding of it as a form of communicative action in digital culture. Genre is a means of understanding typified forms of social and communicative actions in particular forms of human activity spheres. Specifically, I want to underpin how we can begin to understand the organization of knowledge as a form of typified social and communicative action in digital media.

Second, I am going to consider how the organization of knowledge can be a means to understand features of digital culture. In digital culture, we encounter the activity and practice of ordering and archiving, particularly through search engines, algorithms, and databases, and I am going to look at what they do communicatively in digital culture. Such an understanding has the potential to enhance conversations about new media and the new communicative infrastructures in society, where the activity and practice of the organization of knowledge has increasing social and cultural significance.

The chapter unfolds in the following way. I begin with some brief remarks on the everyday organization of knowledge. For this purpose, I draw on the work of Beer (2013) and his ideas about the archive in digital culture. Next, I consider how genre can be used to explore the understanding of the organization of knowledge suggested in this chapter. By looking into some work done on search engines, databases, and algorithms as a means of communication, I begin to establish the premise from where to argue for the organization of knowledge as a communicative genre in digital culture and as an analytical concept for understanding features of digital culture.

1.2. The Everyday Organization of Knowledge and Communication in Digital Culture

Libraries, archives, and similar kinds of collecting institutions and systems have for a long time been used to share and facilitate access to their particular collections. The whole idea of having collaborating local, regional, national, and global bibliographical systems rests on the assumption that providing access to structured collections of recorded knowledge is a social responsibility and of benefit to particular user communities: the public and societies at large. These efforts continue with digital networks and digital media and play a significant part in society's communicative and information infrastructure.

However, what I would like to pay attention to in this chapter is the communication and organization of knowledge in digital networks that takes place outside these institutions and the practices and activities of their professionals. I will look at the everyday organization of knowledge in the everyday life of people when involved in different communicative interactions, whether that is on Facebook, YouTube, Twitter, Flickr, Wikipedia, or on search engines. In digital culture there is a public expectation to be able to search and to store — and to be searched for and stored — in structured collections of items, because digital media, as opposed to classic mass media, offers, among other things, this exact possibility due to their inherent database configurations (Manovich, 2001), and because of digital media's integration in almost all other forms of media we encounter in our everyday digital culture. Thus, with digital media it seems like digital culture is centered on the idea of organizing, archiving, and ordering in everyday life.

This is also the central theme of David Beer's book, *Popular Culture and New Media* (Beer, 2013). In particular, Beer pays attention to the role of classification in everyday life given the permeation of new media. He labels this as "the classificatory imagination" and argues that "... we need to begin to factor in the decentralization of cultural classification and archiving processes to understand the ordering of culture. In other words we need to develop a stronger sense of the *classificatory imagination* in culture" (Beer, 2013, p. 44; italics in original), with "classificatory imagination" covering the idea that people come to think and act by means of classification in everyday encounters with digital media. This argument about the necessity of factoring in the decentralization and domestication of classification and archiving in everyday is important for the argument to be developed in this chapter. Recognizing that classification and archiving are a means of ordering culture offers an opportunity to reflect upon what this means to the practice and activity of the organization of knowledge in everyday life.

Continuing this line of reasoning, Beer proposes the concept of archives as a means to understand digital culture: "The concept of archives can be used to explore the ordering of data in digital culture and to ask who controls them, what is stored, how it is accessed, how it is managed and so on" (Beer, 2013, p. 41). It is not a concept developed strictly from archival theory. It is used in a rather symbolic way to underscore the archival processes triggered by the proliferation of data in digital culture. Communication in digital networks is facilitated by data produced by people's communicative actions in their infrastructures of participation (Beer, 2013, p. 53). Data is collected by services and platforms and stored in archives, producing traces of our communicative affairs. Digital networks and the communicative actions they enable go hand-in-hand with forms of knowledge organization, or archiving as Beer (2013) argues, as much of the work of categorization, tagging, and ordering is carried out by people in their everyday interactions in digital networks. They do this because they communicate and relate to each other and make sense of society at large by these means. Thus, the concept of archives as an analytical tool offers a way to understand communication in digital culture and digital networks.

Here, I want to take Beer's argument a step further. People and institutions do not communicate and produce communicative actions coincidentally. In order to make sense of, for instance, tagging or archiving as modes of communication and ordering, humans construct structured ways of communication in order to make their communications recognizable and intelligible to their appropriate local audiences and activity contexts. Typified forms of communication and genres are developed.

1.3. Genre: Understanding Local Communicative Interactions and Social Structure

In rhetorically oriented understandings of genre, genre is a typified form of communication employed by people in various spheres of human activity and social practices (Bakhtin, 1986; Bazerman, 1988; Berkenkotter & Huckin, 1995; Devitt, 2004; Luckmann, 2009; Miller, 1984; Yates, 1989). Genre as an analytical concept is a powerful means of understanding diverse forms of communication as it sheds light on not only the very concrete form(s) of text(s) used in human activity spheres, but also on the social structures producing genres and their role in coordinating social organization (Luckmann, 2009). As such, genre is always caught between its broader scope of production and its concrete local context of appropriation and interpretation. The uptake of the genre (Freadman, 1994) in a concrete setting is crucial for its success in coordinating social action. Given that the action of a genre can be caused by activity contexts outside of the particular genre, genre always mediates between a local and global level. For instance, the genre of the scholarly journal article is mediating between its local contexts of production and its extension into multiple disciplinary and globally dispersed contexts. However, the very appropriation of a genre is of course locally dependent. Genre is thus always caught in the tensions and contradictions between local and global communication contexts and, at the same time, is the communicative means available to people for producing recognizable communicative actions between or within local and global contexts.

Genre is a means for making sense of communication at a distance and for making communication recognizable to others. Genre entails specific forms of communicative relationships and positionings among individuals. Typification, coming from Schutz's phenomenology (Schutz, 1967), can help establish and maintain a communicative order at a distance that helps writer and reader share a recognizable space of meaning and make available a common ground for meaningful construction and for meaningful discourse.

Because distance means a distance between something, communication is always mediated. It is not and cannot be pure dialogue between separated minds (Peters, 1999). Genre is always about the other or the addressee, as Bakhtin notes (Bakhtin, 1986). Genre implies the other as genre is a social game and not a solipsistic trap. The other is necessary. Otherwise, distance is not distance. Thus, genre is not a technical fix for communication. It is a communicative solution to socio-communicative problems (Luckmann, 2009). It is a social and cultural tool loaded with the ideologies, motives, ideas, and knowledge of particular groups of people in particular contexts. Genre is not the outcome of some master plan. Genre is pragmatically molded by tensions and contradictions between participants in human activities.

Framing my further observations within this view of genre, I am now going to examine how search engines, algorithms, and databases communicatively perform in digital culture. Having done that, I will move on to elaborate my argument about the organization of knowledge as a concept for making sense of features of digital culture and as a genre in digital culture.

1.4. Search Engines and Communicative Actions

In current media culture, searching and search engines are significant forms of media interaction. On a par with television, radio, and the press, search engines are a new powerful form of mass media which penetrate our everyday lives and public spheres in almost every dimension. "As our use of digital media converges, mixing and combining computer applications with more traditional media, we also find search engines becoming part of our entire media ecosystem" (Halavais, 2009, p. 10). The homepages of public authorities and private companies present themselves as ordered places to be searched in order for us to produce meaningful action with them. Cell phones, apps, and e-books include a search function, and the big search engines such as Google or Bing constantly make us aware of the importance of being searchable, represented, and visible (or invisible) in their collection of digital resources. In other words, search engines have introduced the verb "to search" constituting a new communicative action different from but complementing the verbs attached to traditional mass media such as "to listen, to read, or to watch." Contemporarily, these four verbs and actions represent prominent ways of interacting with various media and cultural forms.

Search engines have induced "a culture of search" (Hillis, Petit, & Jarret, 2013). They help produce the public expectation that almost everything relevant is being recorded and is available and searchable; that everything is coded and archived for retrieval. "Search is a way of life" and "a public utility," as noted by Hillis et al. (2013, pp. 4–5). Although search engines may be a public utility, they are, of course, not neutral, just like all other media. The production of search results does not come from nowhere; it is generated by the deliberate actions of search algorithm designers and those (big companies, for instance) who produce content to be included in search engines.

In addition, searching using search engines can also be seen as a way of connecting with other people and with the world at large: "Every tag, search entry, and click is, in fact, a way of connecting to other people's searches and tags; i.e., to other people's intentions of naming and identifying the things that matter to them most" (Hirsu, 2015, p. 32). Hirsu's description of search reminds us that humans are social creatures and that the desire to connect with other like-minded persons is still with us in the age of search. The difference compared with older forms of media is the way in which we connect (i.e., by searching) to others. Halavais (2009, p. 161) calls this sociable search: "The emergence of sociable search suggests that we need to find not just information, but each other. Search is important for the individual who wants to be part of a larger social conversation ...." Together these observations suggest how search engines perform as a means of communication by focusing on the idea of being connected to others and being part of various social conversations. It further suggests how search engines align with established mass media, as they too rely on the idea of connecting people through conversations about social and cultural matters. In short, search engines inscribe themselves in the history of communication and our hopes of being connected or related to the other (Peters, 1999).

Search engines are a first step toward grasping search as a genre. Search in digital culture can be considered part of a communicative solution to communicative projects (Luckmann, 2009). Search becomes a rather typified form of action engaged in by people when employing search engines. For instance, because of our previous experiences with search engines, they make us aware of how to limit our search parameters, how to phrase our query appropriately, how to produce an adequate search as a response to a particular situation, and, not least, how to align with — or not align with — the search algorithms.

1.5. Algorithms: Between Communication and Culture

Digital media is, among other things, made up of software and with software comes algorithms (Manovich, 2013). Algorithms, or algorithmic communications, have during the last decade shaped our ideas of communication and culture. There is a fast-growing body of literature on the topic (Beer, 2009, 2013; Bucher, 2012, 2017; Cheney-Lippold, 2011; Gillespie, 2014; Hallinan & Striphas, 2016; Kitchin, 2017; Natoli, 2014; Striphas, 2015) pointing to the relevance and urgency of understanding the force and impact of algorithms in current forms of digital culture. Algorithms shape our digital interactions with the very data we produce. They form our ideas of what can be searched for and known. Algorithms shape our online identity, potentially making us unknown to ourselves and challenging our own conception of who we are. Being inscribed in digital media, algorithms at the same time determine what is index-able, what there is to be communicated about, and in what ways. Algorithms do not do anything in human communication unless we, as humans, animate them for a specific purpose in our efforts to connect with others. Algorithms are, of course, also designed and encoded with certain motives, ideologies, or beliefs on behalf of those who put them into action.

Being a part of the regularization of inscription and calculation, algorithms make representations of phenomena similar and interpretable across situations and interlocutors, by treating them as the same. They also enlist those phenomena and the regularized data in regularized activity systems and actions, though they are also open to modification and alternative uses.

Although the rule-governed aspect of algorithms (and particularly the way they are embodied in software) limits their flexibility, we might also think of them as software objects, allowing them to be embedded in various infrastructures. In these infrastructures, algorithms take on agency as they develop "... their own capacity to remember" (Beer, 2013, p. 15) because they are in use. Being used and displaying agency, means they capture aspects of their use (Beer, 2013, p. 17). Therefore, algorithms are "active in capturing aspects of our everyday lives and cultural engagements" (Beer, 2013, p. 20), and we should recognize as the active agents they are.

Although algorithms and the communications they produce lack transparency because we do not know exactly what animates their action besides our own, the transparency issue may not be the problem for understanding their role in digital culture. Rather, it might be beneficial to recognize their agency as something that must be acknowledged as part of culture. For this reason, Striphas (2015) coined the term "algorithmic culture" by which he means "... the use of computational processes to sort, classify, and hierarchize people, places, objects, and ideas" (Striphas, 2015, p. 395). Computers and computational processes are not only pervasive; they are also beginning to influence what we understand as culture because of computational processes, thereby challenging established notions of culture formed in a non-digital world and by people traditionally inhabiting the cultural sphere. By now, algorithms are embedded in our communicative infrastructures and their form of agency helps produce and typify our cultural and social assumptions. But algorithms need data and with data comes databases.

1.6. Databases: The Ordering of Culture and Society

Databases are underneath every sort of new digital media and order the world according to their logic of listing items in a structured collection (Manovich, 2001). Databases and the activity of listing are, of course, not a new activity. Precursors such as bibliographies, lists, or oral recitals have been with us throughout human history and have underpinned many social and cognitive activities. Jack Goody argued most forcefully about the introduction of the list in written cultures, saying that lists "... were not simply by-products of the interaction between writing and say, the economy, filling in some hitherto hidden "need," but that they represented a significant change not only in the nature of transactions, but also in the modes of thought that accompanied them ..." (Goody, 1977, p. 81). Likewise, we may say that the database also represents a change in the mode of thought that accompanies it. A database understands the world according to how a computer understands it (Hayles, 2012, p. 176), that is, in terms of automation, calculation, and symbols. The more widespread databases become in our cultural and social interactions, the more we need to, or are forced to, align with their way of looking at the world. The potential change in mode of thought that databases introduce is a sense of calculated communication mediated through structured collections of items, whether that is in cell phones, social media, search engines, or mass media; socially and culturally relevant pieces of information are amassed in structured collections whose main mode of communication is search and retrieval.

Arguing that the database is a new cultural form, Manovich (2001) offers a way of thinking about how it is that databases shape the ordering of culture and communicative actions. This idea about the database is not to say that it stops being a technical object. Rather, what Manovich points to is that with new (digital) media being used in cultural production and consumption, the database and its logic of ordering moves into the cultural sphere. We may of course question whether this argument applies to all forms of new media, as Manovich was writing at a time when print materials were beginning to be digitized and, hence, moved into a database. Today, we have digital-born materials which differ from digitized materials coming from non-digital material (Brügger & Finnemann, 2013). Nevertheless, the database as a cultural form is both a useful metaphor and concrete tool for understanding the role of ordering in contemporary culture. For instance, Google, Twitter, or Facebook position themselves as being of great benefit to human social interaction. However, they are also players on a competitive media market and their capital is, among other things, the data they collect in their databases about our online activities.

Moreover, the database and its mode of being have moved into culture and our consciousness. Of course it does not completely occupy culture and our consciousness, but the database has gained a presence to an extent we may not have witnessed before because of its integration in many forms of digital media. Its presence has resulted in the ideas of data, search, retrieval, and storage having a strong manifestation in culture and in our everyday practices. Thus, like Beer's concept of the archive, the database as a cultural form in its own right provides a means to understand current cultural production and articulations through its way of seeing the world, that is, through data and categories of data.

1.7. The Organization of Knowledge as Analytical Concept in Digital Culture

Partly influenced by Beer's idea of the concept of archiving as a means to understand digital culture (Beer, 2013), the organization of knowledge makes us see digital culture as a culture of ordering, organizing, or archiving items and how this is inscribed in culture. Obviously, this is not a whole new feature of culture and cultural production. But the organization of knowledge is a new cultural form (cf. Andersen, 2008; Manovich, 2001) in the sense that it is not foreign to people nowadays in their everyday lives to participate in, use, or be constituted by forms of archiving. People may not call it by that name, but they essentially carry out the practice of organizing or archiving knowledge when making their way around digital media culture.

The organization of knowledge forces us to make sense of our everyday world through how categories and categorizations are produced because that is how many digital resources and platforms speak to us. That is, we have a conversation partner, whether that is a database, a search engine, or algorithms who speak through the logic of structuring, numbering, and ordering but whose communicative actions are dependent on ours.

The database speaks through its structured collections and the communicative action it engenders, namely, search and retrieval. Persons included in a database means being identified and given an identity (Poster, 1995) according to the fit with the structures and categories of the database and its producers. As with all structured collections of items, there is a duality between who determines the categories and what to put in them, and who looks at the categories and potentially makes sense of them. Databases and their contents are products of larger communicative projects (Luckmann, 2009) to which they must fit in. As such, databases can be considered a genre as they are communicative solutions typified through the categories they communicate.

The search engine as a partner in conversation provides the hope of being connected to the "other" through clicks, search histories, or keywords. It provides the expectation that most things (shoes, events, food, clothes, places, buildings, streets, etc.) are coded for search and what cannot be found does not exist. Search engines thus perform differently than libraries, as their audiences use them for different activities — activities libraries cannot (and maybe should not) fulfill. Due to their historical origins and their connection to the written word, libraries symbolize a literary culture. Search engines do not symbolize a literary culture. Search engines reflect the culture of the market. For instance, companies who want to be visible and searchable in search engines most likely do not conceive of search engines as part of literary culture, but rather as a marketplace for advertising.

Thus, as an analytical concept, the organization of knowledge offers a way to make sense of current forms of digital culture and social interactions. This means that we as people and as citizens become more engaged with digital media because of their omnipresence in our life, we also begin to understand and talk about our social and cultural interactions in terms of what, how, and who can be tagged, archived, or searched for. Tagging, archiving, or searching as social and cultural interactions assume some sort of structured collections of items. They, moreover, assume — albeit implicitly — some sort of activity (algorithmic or human) producing those collections. The ordinary presence of the phrases "I googled ..., " "I tweeted ..., " or "I tagged ..." suggests how we relate to and connect with others through communicative actions enabled by digital media and their structured collections of items. To google, to tweet, or to tag reveals a consciousness of ourselves as actively involved agents in communication, but these also reveal a habitualized form of human action shaped by the affordances of search engines and social media in our lives.

To understand our everyday interactions, our cultural assumptions, or our social worlds through the concept of the organization of knowledge is a way of recognizing the permeation of the powers of organizing in our infrastructures of participation (Beer, 2013). The concept of the organization of knowledge helps to make clear what we may already know as individuals in digital culture: that we are configured by and inscribed in categories and sets of data that we ourselves have partly produced by means of our digital communicative actions.

The concept brings to our attention the social and cultural force of categories and how difficult it can be to get out of them once you are included and thusly looked upon by others. Accordingly, the concept of the organization of knowledge reminds us — as have thinkers like Marx, Foucault or Bourdieu — of the ever, ongoing power struggles between classes, between cultures, or between those who categorize and those who are categorized. This reminder may have more pertinence than ever given the saturation of digitally structured collection items in our everyday lives.

1.8. The Organization of Knowledge as a Genre in Digital Culture

In analyzing blogging as a genre, Miller and Shepherd (2004, p. 1) contended that "When a type of discourse or communicative action acquires a common name within a given context or community, that's a good sign that it's functioning as a genre." In our current form of digital culture, to tag or to google are clearly common names for a type of communicative action performed by people in particular everyday contexts with particular forms of digital media. Looking at the organization of knowledge (i.e., tagging, organizing, ordering, archiving, or searching) as a genre in digital culture is a way of understanding it as a tool in communication, helping to typify, stabilize, and make socially recognizable different forms of communicative actions. In digital culture, it is something we make use of in order to make sense of situations and to be able to act appropriately in given situations demanding our responses. For example, with regard to tagging, Beer (2013, pp. 53–54) claimed that this is "... one of the most powerful organizing and ordering practices in contemporary culture ... central in understanding the organization and ordering of contemporary culture. It shapes cultural encounters, it defines how archives work, and what is retrievable ..." Beer's observation is useful in this context as he connects the practice of tagging with an understanding of contemporary culture. Extending this observation to tagging understood as a specific form of typified communication, we might begin to see how it is that tagging as "organizing and ordering practices" are powerful in contemporary culture. For instance, when one chooses to tag a particular item with the tags X, Y, and Z, these tags also begin to typify what is to be communicated with them. They are "stabilized-for-now," as Schryer (1993) says about genre. The tags create expectations as to what is to be found within and communicated them: "Once one has 'chosen' a genre for a communicative project, it is the genre that 'chooses' the parts for its accomplishments" (Luckmann, 2009, p. 273). Twitter can serve as an example of Luckmann's notion of genre in communicative projects. The hashtag, as a knowledge organization component of tweeting, can be used for a variety of functions. Sometimes, hashtags are used to facilitate retrieval of tweets on similar topics or associated with particular events, and to enable systematic, conscious collocation of a set of similar tweets: people at a conference might be encouraged to use a particular hashtag so that all the tweets from that conference can be found together (#ECREA2016), or a hashtag might collocate tweets that comment on a particular issue (#cyberbullying) or event (#rio2016 for the Olympic Games). But tweets are used for much more than this. A hashtag like "#blacklivesmatter" is not used just to facilitate sorting and retrieval but to declare one's allegiances and invoke a particular perspective. In other words, here the "tag" is performing an explicit communicative function — and, importantly, use of the tag is recognized as such by authors and readers of tweets — in addition to facilitating retrieval and sorting. In other words, hashtags in Twitter work somewhat differently than other tags, such as those associated with photos in image services, and searching in Twitter is also somewhat different than searching images. Twitter hashtags are (at least sometimes) read more like a newspaper or a poem than like a temperature reading on a thermometer.

To take Luckmann's argument to its logical conclusion, we can say that to communicate by means of a chosen tag, and for the tag to accomplish a communicative effect in a communicative project, the chosen tag commits us as communicators. The tag or tags need to be socially recognized and accepted. Thus, tagging in contemporary digital culture is not only a "powerful organizing practice" because it shapes "cultural encounters." It is powerful as an organizing practice because it gives rise to a typified form of communication; that is, as genre. Genres are powerful communicative tools because they shape communicative interactions and commit people in these interactions in structured ways (Paré, 2002).

1.9. Conclusion

What I have argued in this chapter is that in digital culture we can perceive the organization of knowledge as both an analytical concept to understand communication in digital culture and as a particular genre in digital communication. The implication of this argument is, first, that the organization of knowledge can be considered a keyword in our cultural vocabulary (Williams, 1983), making it part of the language by which we understand and describe digital culture. Having the status of a keyword in digital culture shapes an understanding of the organization of knowledge as moving from being limited to a particular professional practice to a way of perceiving a range of features in digital culture having to do with, for instance, data archives, data circulation, digital cultural production, and consumption or algorithmic cultures (Striphas, 2015).

Considering the organization of knowledge as a genre in digital communication is one way of suggesting its social function as a tool and not as an objective to be achieved in human communication (Andersen, 2015). Being a genre in digital everyday communication suggests that the organization of knowledge is one genre among many in digital culture. But it is a particular genre that has come about because of the architecture of digital media as it invites ordering practices due to its databased nature. As more and more people on a daily basis are in touch with, or are implicated by, structured collections of items as part of their communicative actions, this points to the potential in seeing the organization of knowledge as a genre. It ascribes to the practices of ordering and archiving a form of agency in digital culture.

References

Andersen, J. (2008). Knowledge organization as a cultural form: From knowledge organization to knowledge design. In J. T. Tennis & C. Arsenault (Eds.), *Proceedings of the tenth international ISKO conference*, Montréal (Vol. 11, pp. 269–274). Würzburg: Ergon Verlag.

Andersen, J. (2015). Re-describing knowledge organization – A genre and activity-based view. In J. Andersen (Ed.), *Genre theory in information studies* (Vol. 11, pp. 13–42). Studies in Information. Bingley: Emerald Group Publishing Limited.

Bakhtin, M. (1986). *The problem of speech genres* (V. W. McGee, Trans.). In C. Emerson & M. Holquist (Eds.), *Speech genres and other late essays* (pp. 60–102). Austin: University of Texas Press.

Bazerman, C. (1988). *Shaping written knowledge. The genre and activity of the experimental article in science*. Madison, WI: University of Wisconsin Press.

Beer, D. (2009). Power through the algorithm? Participatory web cultures and the technological unconscious. *New Media & Society*, *11*(6), 985–1002.

Beer, D. (2013). *Popular culture and new media. The politics of circulation*. London: Palgrave Macmillan.

Berkenkotter, C., & Huckin, T. (1995). *Genre knowledge in disciplinary communication: Cognition/culture/power*. Hillsdale, NJ: L. Erlbaum Associates.

Brügger, N., & Finnemann, N. O. (2013). The web and digital humanities: Theoretical and methodological concerns. *Journal of Broadcasting and Electronic Media*, *57*(1), 66–80. doi:10.1080/08838151.2012.761699

Bucher, T. (2012). Want to be on the top? Algorithmic power and the threat of invisibility on Facebook. *New Media & Society*, *14*(7), 1164–1180.

Bucher, T. (2017). The algorithmic imaginary: Exploring the ordinary affects of Facebook algorithms. *Information, Communication & Society*, *20*(1), 30–44.

Cheney-Lippold, J. (2011). A new algorithmic identity. *Theory, Culture & Society*, *28*(6), 164–181.

Devitt, A. J. (2004). *Writing genres*. Carbondale, IL: Southern Illinois University Press.

Freadman, A. (1994). Anyone for tennis? In A. Freedman & P. Medway (Eds.), *Genre and the New Rhetoric* (pp. 43–66). London: Taylor & Francis.

Gillespie, T. (2014). The relevance of algorithms. In T. Gillespie, P. J. Boczkowski, & K. A. Foot (Eds.), *Media technologies. Essays on communication, materiality, and society* (pp. 167–193). Cambridge, MA: MIT Press.

Goody, J. (1977). *The domestication of the savage mind*. Cambridge: Cambridge University Press.

Halavais, A. (2009). *Search engine society*. Cambridge: Polity Press.

Hallinan, B., & Striphas, T. (2016). Recommended for you: The Netflix Prize and the production of algorithmic culture. *New Media & Society*, *18*(1), 117–137.

Hayles, N. K. (2012). *How we think. Digital media and contemporary technogenesis*. Chicago, IL: The University of Chicago Press.

Hillis, K., Petit, M., & Jarret, K. (2013). *Google and the culture of search*. New York, NY: Routledge.

Hirsu, L. (2015). Tag writing, search engines and cultural scripts. *Computers and Composition*, *35*, 30–40.

Kitchin, R. (2017). Thinking critically about algorithms. *Information, Communication & Society*, *20*(1), 14–29.

Luckmann, T. (2009). Observations on the structure and function of communicative genres. *Semiotica*, *173*(1–4), 267–282.

Manovich, L. (2001). *The language of new media*. Cambridge, MA: MIT Press.

Manovich, L. (2013). *Software takes command*. New York, NY: Bloomsbury.

Miller, C. R. (1984). Genre as social action. *Quarterly Journal of Speech*, *70*, 151–167.

Miller, C. R., & Shepherd, D. (2004). Blogging as social action: A genre analysis of the weblog. *Into the Blogosphere: Rhetoric, Community, and Culture of Weblogs*. Retrieved from http://works.bepress.com/dawn_shepherd/2/

Natoli, P. M. (2014). Automated media: An institutional theory perspective on algorithmic media production and consumption. *Communication Theory*, *24*, 340–360.

Paré, A. (2002). Genre and identity: Individuals, institutions, and ideology. In R. M. Coe, L. Lingard, & T. Teslenko (Eds.), *The rhetoric and ideology of genre: Strategies for stability and change* (pp. 57–71). Cresskill, NJ: Hampton Press.

Peters, J. D. (1999). *Speaking into the air: A history of the idea of communication*. Chicago, IL: University of Chicago Press.

Poster, M. (1995). Databases as discourse, or electronic interpellations. In M. Poster (Ed.), *The second media age* (pp. 78–94). Cambridge: Polity Press.

Schryer, C. F. (1993). Records as genre. *Written Communication*, *10*, 200–234.

Schutz, A. (1967). *The phenomenology of the social world*. Evanston, IL: Northwestern University Press.

Striphas, T. (2015). Algorithmic culture. *European Journal of Cultural Studies*, *18*(4–5), 395–412.

Williams, R. (1983). *Keywords: A vocabulary of culture and society* (Rev. ed.). New York, NY: Oxford University Press.

Williams, R. (1989). Culture is ordinary. In *Resources of hope: Culture, democracy, socialism* (pp. 3–18). London: Verso.

Yates, J. (1989). *Control through communication: The rise of system in American management*. Baltimore, MD: Johns Hopkins University Press.

Chapter 2

Information Cultures: Shapes and Shapings of Information

Laura Skouvig

Abstract

The purpose of this chapter is to suggest a genealogy of the concept of information beyond the 20th century. The chapter discusses how the concept of information culture might provide a way of formulating such a genealogic strategy. The chapter approaches this purpose by providing a general narrative of premodern information cultures, examining works on early-modern scholars and 18th century savants and discussion of what seems to be a Foucauldian rupture in the conceptualization of information in 19th century England. The findings of the chapter are situated in the thinking that a genealogy of information would reveal that information had specific purposes in specific settings.

Keywords: Information history; information culture; Foucault; Denmark 1800–1815

Chapter Outline

The Organization of Knowledge: Caught Between Global Structures and Local Meaning
Studies in Information, Volume 12, 17–31

ISSN: 2055-5377/doi:10.1108/S2055-537720170000011008

Much literature has dealt with the present information society questioning the meaning of the term information. As a key marker of the present social formation, getting a grip on this term should be welcome. However, it seems that information is a rather elusive idea that tends to escape any precise definition (Black, 2006, p. 442). The vagueness surrounding the concept of information makes it problematic as a key identifier of the information society (Webster, 2006, p. 444). The British sociologist, Frank Webster, addresses this difficulty and argues that conceptions of the information society depend on quantity as a defining criterion (Webster, 2006, p. 444). Stating that the information society is defined by the presence of more information seems inadequate, particularly since it does not provide any guidance on interpreting the meaning of information itself. In addition to quantity as a signifier, Webster also points to the pervasiveness of technologies as an important part of the definition of the present information society. Information technologies enable people to enjoy smooth and constant access to information. Accessibility becomes a cornerstone in the well-informed democracy (Webster, 2014, p. 214). This understanding of information often includes a progressive perception of technology which provides a distinct framing of the concept of information and the conceptualization of information society (Webster, 2006, p. 445). Webster argues for the need to "... qualify the generalizing terminology ..." (Webster, 2014, p. 1) though he acknowledges that such terms also include possible paths to understanding the present society. How then to qualify the terms information and information society? Webster has, in an earlier article, called for a Foucauldian genealogy that scrutinizes how the "blanket term" (Webster, 2006, p. 452) is filled out and understood by those who frequently use the term information, and thus looks at information as a product of discourse (Webster, 2006, p. 452). Webster seems to extend the discourse of information and genealogy as a strategy only to 20th century discussions about information. I argue here for a separation of the two Foucauldian strategies of analysis and for a focus on genealogy, but a genealogy that challenges Webster's 20th century boundary.

This is a bold undertaking for an article, so my aim here is to discuss how the concept of information culture might provide a way of formulating a genealogic strategy. This chapter provides a general narrative of early modern information cultures, examining works on early-modern scholars and 18th century savants and discussion of what seems to be a Foucauldian rupture in the conceptualization of information in 19th century England. This section is followed by a genealogical reading of the period of late absolutist rule in Denmark, focusing on how information was produced, organized, understood, and distributed in a culture of information secrecy. From a Foucauldian perspective, historical investigations provide unique possibilities for unraveling and questioning transparent and naturalized conceptions of contemporary society. Foucault argues in the essay *Nietzsche, Genealogy, History* that historians blindly accept their present point of view and regard history as a continuing all-encompassing progression (Foucault, 1984a). Rather, historians should use history to critically illuminate the emergence of present conceptions. In Foucault's thinking, history is part of a philosophical reflection on problem areas in current society and culture (Foucault, 1984b; Raffnsøe, Gudmand-Høyer, & Thaning, 2008). By using Foucault's genealogy, this chapter aims at questioning the present understanding of information by "... disclos[ing] dispersions and differences, [...] leav[ing] things undisturbed in their own dimension and intensity" (Foucault, 1984a, p. 89). History becomes a heterotopia (Foucault, 2002) of the present — a space that reflects and twists the present — making us ask new questions and take new actions. Genealogy is in short about unraveling present practices and locating emergences without presenting a chronology of progression.

Foucault stresses that the genealogy is "... gray, meticulous and patiently documentary" (Foucault, 1984a, p.76) and relies on concrete research of past cultures and practices. In many ways, one could read an essay by media historian John Durham Peters as a genealogy of information itself. In this essay, he traces the changes in information as a concept as well as the practices around it from its first appearance in medieval English to its form after World War II. He examines how the medieval understanding implied a meaning of giving form and shape. This understanding of information as formative turned into a notion of information as stuff and matter through 17th century empiricist science. Information became something that made imprints in the soft wax of the human mind (Peters, 1988, p. 12). The main argument in Peters, analysis is that information over time became externalized and objectified. This was emphasized by the replacement of the individual human being as the only knower by the state bureaucracy as a common knower. Statistics became valid information and bureaucrats sought techniques to handle and control the quantitatively growing amount of information. State bureaucracy gave birth to the computer as a primary information processing machine and the final reification of information (Peters, 1988, p. 16).

2.1. Unraveling Notions of Information Cultures

Information culture is an analytical term that enables me to shed a light on historical periods and gain new insights when looking into how people understood and used information in the past. The British information historian, Toni Weller, defines information culture as: "... cultural mediums of information dissemination and exchange but also as influencing culture through a broader sense of information as an increasingly important and recognized part of society" (Weller, 2011, p. 200). A strong impetus to the development of the concept of information culture is the insight that information has no fixed meaning but is dependent on its historical, cultural, social, technological, and media contexts. Information is not a thing with a fixed meaning that transcends centuries and cultures. It is fluent, contingent, and interchangeable, as argued by Peters (1988). It is a strong marker of the present information culture that information is recognized and verbalized as a distinct and important feature. From a genealogical point of view, this is exactly what should be questioned. Studies of past information cultures need not focus on quantities or qualities of information but simply take as a starting point that all societies are information societies (Darnton, 2010; Finnemann, 2005), in order to unravel how information appeared, which forms it took, and how people perceived it. From my point of view, information culture also implies a close affinity to the material conditions of information production and use (Knorr Cetina, 2007).

The American book historian, Ann Blair, argues that early-modern scholars practiced "... storing, retrieving, selecting, and organizing information, with the implication that it can be stored and shared for use and reuse in different ways by many people ..." (Blair, 2010, p. 3). She identifies behind these practices a distinguishing factor which she calls "info lust" (Blair, 2010, p. 6). This info lust thrived together with a sense of information overload evoked by changes in the quantity and quality of information, the existing tools for handling information, and cultural and personal expectations (Blair, 2010, p. 3). All these different trends had to be managed by the individual scholar and by the scholarly community in general. The scholars in Renaissance Europe redefined, reinterpreted, and conceived of new information management tools — particularly reference tools, many of which still exist today as markers of our present information culture (Blair, 2010, p. 265). Blair depicts an information culture of learned scholars who had a keen interest in gathering, systematizing, and disseminating information because they feared losing it.

Blair wants to contest the traditional interpretation of the printing press as the major agent of change between medieval and early-modern information cultures. In her interpretation, the technological powers of the printing press were merely one aspect of a series of multifaceted social events that brought about this sense of overload. Though the printing press produced vast numbers of books, the printed book also became a means for controlling the overload. Blair seems to argue that changes between information cultures cannot alone be explained by new technologies or media — or at least that the role of technology was more nuanced than that. Simply to explain change as a product of new technologies would imply a technological determinism that credits technology as a revolutionizing force and consequently fails to consider technology as a product of human imagination and use (Finnemann, 2005, pp. 42–43).

In Blair's work, conceptions of information become more influential in an information culture than the available technology. It is these conceptions that lead to new notions and uses of technology. The technology was used to frame and handle the needs that emerged from the specific understanding of information among the learned culture of scholars in early modern Europe.

Within this line of thinking, the American media theorist, Neil Postman, argues that we today do not conceive of the Age of Enlightenment as an information age simply because people then had a different notion of information (Postman, 1999, p. 85). Even though a 21st century view of the 18th century recognizes how it was characterized by an unprecedented stream of information and new institutions and media that produced and communicated information, the word itself rarely appeared. Information was nothing in itself and had no separate existence (Postman, 1999, pp. 82–83). Up to the middle of the 19th century, information was understood rhetorically and always embedded in a context giving it a purpose. For instance, Postman argues, Diderot (one of the initiators of the French Encyclopédie) saw information as a weapon in the fight for societal change through applied skepticism (Postman, 1999, p. 86). The French encyclopedia was thus not just a mere commodity and provider of, or a vessel for, information, abstracted and applicable in a variety of situations, but contained information aimed at changing the world. Information "… gave shape, texture, or authority to a political, social, or scientific concept" (Postman, 1999, p. 86). Information should *do* something — preferably, make people change the world.

From this perspective, Postman argues that the notion of information as abstract and context-free, with a value on its own, did not exist prior to the advent of telegraphy and photography in the middle of the 19th century. Information as a commodity was a feature of modern society. However, even the encyclopedia was for sale and Postman ignores the fact that information, particularly in the form of news, has always been for sale (Burke, 2000, pp. 149–177). Postman and Blair both argue that attitudes toward information are profound markers of distinct information cultures, but where Blair recoils from pointing at technology as a reason for change, Postman suggests that technological changes resulted in a change in the conception of information toward its present definition (Postman, 1999, pp. 87–89).

It should not be questioned here whether the advent of telegraphy and photography had an impact on how information was conceived, but these technologies were mixed up in a broader social and cultural transformation, as emphasized by Weller (2009). These other cultural and societal changes contributed to this new, modern conception of information as abstracted, neutral, and objective. In that form, information became something people were able to claim a right to (Weller, 2009, p. 165). In a study of information perceptions in the United Kingdom in the 19th century, she argues that the changing view of information stemmed from the general need for social control in a society marked by strong industrialization. Her analysis shows that popular media helped shape information as a cultural commodity, but it was a commodity that thrived in the open and public sphere. Over the course of the 19th century, information lost its rhetorical framing and context (Weller, 2009, p. 56). Information turned into a public good and became finally an end in itself.

This short discussion of aspects of early modern information cultures illuminates most importantly how information itself was understood differently and how this resulted in different views on technology and quantity of information. Peters, Postman, and Weller are interested in understanding how information became this externalized, abstracted context-free *thing* in the (late) modern information age. Together with Blair, Postman and especially Weller frame the role of technology as merely one among several factors that influenced how information was understood and used. In this way, they help question the inter-linkage between technology and progress in history. In addition, Blair's analysis criticizes the question of quantity. The learned community in early modern Europe faced a huge amount of information simply because they wanted to gather and handle it. Her study shows that quantity is rather a matter of perception.

In the next section, I will sketch out how an early modern understanding of information as secret defined how information was used, produced, and communicated at the end of the 18th and the beginning of the 19th century in Denmark.

2.2. An Absolutist Culture of Information: Denmark at the Turn of the 18th Century

Absolutism as the guiding principle of government was adopted in Denmark in 1660. This system of government replaced the elective monarchy based on a strong nobility that had, by then, become considered antiquated. In 1665, the principles of absolutism were recorded in Lex Regia — the only written absolutist constitution in Europe. In its pure form, absolutism meant that all power was conferred on the king. The king ruled by a divine commandment and was only accountable to his own conscience and subject to no-one but God. During the reigns of the absolute kings, there was no division between administration, government, or legislation (Bonderup, 2012). Yet, the king did not rule on his own but relied on a massive, centralized administration and (personal) advisers.

The relations between the king and the civil servants of the central administration and also between the king and his subjects are crucial markers of the space in which information could be managed, understood, and used. During the almost 200 years of absolutist rule, these relations changed. More civil servants were recruited from the rising class of the bourgeoisie and during the 18th century Enlightenment ideas meant that the absolute powers of the king were reinterpreted and seen as a contract between king and subjects (Feldbæk, 2000, p. 242). Civil servants began to advocate for a system of government that relied on public opinion (Feldbæk, 2000, p. 335). This reflected the general attitude that the 18th century Enlightenment was a move toward — or at least demand for — transparency as one of the crucial demands of the bourgeois public sphere (Habermas, 1990). The shift toward an absolutist government based on public opinion is contested within Danish history since the absolutist and divine powers of the king were omitted from critical discussion (Engberg, 2009, p. 282). However, the underlying principle of absolutism remained: the king was the absolute ruler and had no obligation to take advice from either advisors or the public. Information was gathered, produced, and communicated within an understanding of information as something to be guarded, controlled, and not of interest to the general public.

2.2.1. Secret Information and Information Control

The absolutist information culture revolved around two central axes: how bureaucracy performed and understood information management internally and how absolutist rule controlled information in the relation to the subjects. In both cases, a simple principle was the basis for information practices — and that was the concept of secret information. Secrecy was not merely a tool for the governing elite but a general principle in early modern Europe (Smith, 2011, p. 50–51). This understanding of secret knowledge represented a solid contrast to the openness of the French encyclopedia (Smith, 2011, p. 48).

The absolutist system of government facilitated a particular way of governing by means of the central administration and it furthered a distinct space for understanding politics. Politics was not a public affair but rested on the notion of the *arcana imperii* — that governing was based on secrets and exclusion. An analysis of government in the Venetian republic in the 16th and 17th centuries shows how a conception of *arcana imperii* implied a high form of knowledge that essentially was forbidden for the lower classes to obtain (de Vivo, 2007, pp. 2, 86–89). With the *arcana imperii* as the common basis for governing information, a consequence was a strict control and manipulation with information. Published information only reflected the uncontroversial results of complicated and prolonged negotiations in the senate whereas the in-between results were not considered relevant for anyone other than the senators (de Vivo, 2007, p. 19). This kind of information control dealt with the contradictions between public and secret information. It was a reflection of the dilemma of whether to prohibit political information entirely or publish parts of it when necessary. Information and communication were expressions of a particular way of enacting government for Venetian rulers in the 16th and 17th centuries (de Vivo, 2007, p. 4). When the Danish absolutist government communicated with its citizens, the preferred channel was proclamations. These proclamations contained the information that the authorities conceived as relevant for the general public to obtain, for example, to prepare the citizens for war (Skouvig, 2015b). The proclamations were also necessary in order to control information that spread fairly unrestricted through the different social strata in Copenhagen on the streets. This information was not necessarily false rumors — but it was not under control. The government conceived of information as dangerous and something that needed to be governed and controlled, whereas the educated, cultured elite was a proponent of information as free and accessible. However, the cultured elite did also express a conception of information as dangerous for the lower classes in society because they did not possess the same critical and reflective attitude toward information. Thus, information control seemed to be a governing principle in a premodern information culture.

Absolutist government relied on an exclusive and excluding knowledge of foreign policy that it had no intention of sharing with any kind of public. The field of foreign policy had the status of being the exclusive domain of the king alone (Krefting, Nøding, & Ringvej, 2014, p. 62). This had a defining influence on how information was understood in the central administration of absolutist rule. Consequently, news about foreign policy merely consisted of reports of events without clarifying or explanatory comments (Krefting et al., 2014, p. 62). The public had — in this understanding — no interest in politics and should not be involved in foreign policy at all. The public — by Koselleck's definition a mix of old nobility, the rising economic bourgeoisie, and the educated bourgeoisie — expanded its powers economically but was not granted corresponding political influence (Koselleck, 2013, p. 51). This moral paradox concerning economic power and lack of political influence, Koselleck claims, resulted in a fundamental crisis of absolutism and eventually its fall. In the 18th century, the *arcana imperii* represented a glaring contrast to the rising voices calling for public transparency (Krefting et al., 2014, p. 47), but it remained of decisive influence. Information was in general not conceived of as a public good or right. However, as Koselleck demonstrated in his seminal work on the relation between absolutism and the Enlightenment, the notion of secrecy also provided the means for the emergence of critique against the absolutist state in the form of the many secret associations, for example, the freemasons (Koselleck, 2013, p. 68ff).

The understanding of information as secret implied that information control as an aspect of governing was a key area for early modern states (de Vivo, 2007; Giddens, 1985). Information was necessary for the state as an instrument for decision-making (cf. Rule, 1974) and at the same time information could turn into a major problem for the government. Information became a problem when people (the civilized bourgeois as well as the lower classes) came in touch with information out of government control. This made it necessary to control the information of the streets that thrived among the lower classes. Rumors were particularly troublesome since they could lead to the overthrow of the monarchy itself (Darnton, 2010). In Venice, concrete urban spaces marked specific information areas. These areas were defined by varying degrees of access to information (de Vivo, 2007). In Paris, gossip and politics were for sale among ordinary citizens who had no connection to politics or the court (Darnton, 2010). Broadside ballads and the ballad mongers were probable hotspots for the exchange of questionable information in Copenhagen and from 1805 the ballad mongers had the dubious honor of being perceived as deviants. More precisely, broadside ballads were prohibited in 1805 because, as stated by the decree, the information contained in the ballads often involved rumors about private persons or public arrangements (Schou, 1808) which were regarded as dangerous for the public order. The ballad mongers were a noisy and vibrant element in the streets of Copenhagen, singing their stories to the mob. Besides the risk of slander, the broadsides were regarded as trash literature that only sought sensation (Skouvig, 2012, p. 359ff). Focusing on emotions, the primary content of the broadside was a negotiation of interpretations of specific events, for example, navy deserters from the first battle of Copenhagen in 1801 (Skouvig, 2012, p. 359). Broadsides were feared by the government as they were probable conveyers of (untrue) rumors and distrusted by the educated bourgeoisie because they lacked an edifying and educating intention (Skouvig, 2012).

News of foreign policy came with foreign newspapers, letters from abroad, and other news that was brought home by sailors. The newspapers and letters came over the mail routes from Hamburg and arrived twice a week (Skouvig, 2015b). Danish newspapers transmitted European news by provenance, without comment. By provenance meant that the news was registered after country and not, for example, after importance. In accordance with a decree on censorship from 1701, it was prohibited to combine news and opinions, leaving the latter for the emerging and vibrant world of magazines, weeklies, and other periodicals (Krefting et al., 2014, p. 40–41). News in Danish newspaper consisted of reports on the whereabouts of the royal family, matter-of-fact depictions of foreign news, and interminable recordings of who had been admitted into which rank. The strong regulations against providing opinions in newspapers opened a creative space to do just that. Writing became experimental and challenged the authorities (Krefting et al., 2014, p. 42).

Much writing was about political and military events on the European continent during the Napoleonic wars and thus challenged the secrecy of foreign policy. One way of controlling this was simply forbidding newspapers from writing about certain international news, such as the flagrant defeat of Napoleon in Russia. According to Engberg (2009, p. 372), the Russian envoy in Copenhagen found this so outrageous that he set up a reading room with access to foreign newspapers. Engberg does not explain if this opportunity was seized by the citizens in Copenhagen, but the example is interesting as a case of how control led to the bypassing of power and defiance of the authorities (see also Skouvig, 2017). Critique of the absolutist government resulted at best in exile, whereas critique of the general conditions of the country or "grumbling with acrimony" (such a formulation appeared in a decree from 1799 that restricted freedom of expression) over the government could lead to imprisonment (Oldrup, 2007, p. 160). Under heavy interest from the public, the poets P. A. Heiberg and Malthe Conrad Bruun were exiled from the country in 1799 and an ordinary citizen, Michael Brabrand, got into difficulties with the law after his critique of the absolute monarchy (Langen, 2012). Still exempted from public debate were the king as person, absolutist rule and religion, as emphasized in the 1799 decree on the liberty of the press (Engberg, 2009, p. 282).

During the Napoleon wars, increased control and surveillance emerged (Skouvig, 2015a). The horrors of the French revolution were still apparent and aroused some anxiety in the mind of the governing crown prince (Engberg, 2009, p. 282). In this concrete but also perceived state of war, information became even more difficult for a government to obtain and to control. Lacking information, people turned to rumors. Rumors thrived and they were dangerous. The king openly addressed this in 1812, warning the civilized and cultured strata of society against relying on rumor mongers (Royal Open Letter, 1812). Interestingly this open letter also displayed the ambiguity of information control. The cultured elite should distrust rumor mongers and their untrue information, but at the same time the king was well aware that this particular group also corresponded with contacts abroad. Since the country was at war, this correspondence could contain valuable information. By means of the open letter, the king instructed the merchants to report the latter kind of information (Royal Open letter,

1812). It is possible that the royal chancellery lacked information too due to wartime restrictions.

2.2.2. *A Vertical Information Landscape*

The above-mentioned royal open letter reveals one aspect of how the government communicated with its subjects — here primarily in Copenhagen. The letter was distributed at the stock exchange with the superintendent of the stock exchange as mediator. The authorities communicated directly with the subjects through proclamations posted on walls in Copenhagen and read aloud after church services. If subjects wanted to address the king (or his bureaucracy) they had the means of petitions. Communication — and information — was vertical (Krefting et al., 2014, p. 30). The civil servants in the state administration mediated this communication in various respects — but always under the assigned absolute power. During the second battle of Copenhagen in 1807, there was an urgent need for communicating with the citizens in Copenhagen — even for the military authorities. Communication lines were apparently strictly divided between the civil part of the city and the military part since Peymann, the commander-in-chief, had to ask the commissioner of the police in Copenhagen to publish an announcement (drafted in writing). Through the letter to the commissioner of the police, the announcement was referred to the civil sphere (Skouvig, 2015b). This particular announcement (Peymann, September 1, 1807) contained information about the initial refusal to accept the British terms and thus informed citizens about the possibility of direct military engagement. The rhetorical framing underscored the gravity of the situation. The information was prepared and distributed to the people by the military and civil authorities of the Danish absolutist monarchy. Peymann used the well-known communication lines and genres of the absolutist information culture in order to address his fellow citizens and spoke by means of the assigned royal powers (Skouvig, 2015b). The king and his administration conceived of information as their legitimate possession, reducing the role of the population to passive recipients of the selected information and active providers of information to the administration (Krefting et al., 2014, p. 158). The vertical information landscape also meant that the central administration relied on access to information about citizens, their affairs, and the state of the country in order to improve trade, industry, and agriculture. Yet it is important to acknowledge that information had a distinct purpose: whenever it was produced, gathered, disseminated, or controlled it was to the benefit of the absolutist state.

This impression of the absolutist information culture is rather bleak. In a glimpse, we are provided with an understanding of information as something that should be suppressed — or at least controlled — to the benefit of king and state. The thought of information as something in free flow was not conceived of by the absolutist government. This bleak and critical picture should be accompanied by another interpretation of late absolutism. In this interpretation, the king's administration and civil servants strived to obtain the happiness for all subjects. This does not, however, contradict the understanding of information as something to be controlled and disseminated only after it had been thoroughly sifted and evaluated.

The tight control of information during absolutism obviously showed that information was not a public good and that subjects did not have the right to obtain information. The fear of rumors during the Napoleonic wars also showed that the need for information grew where there was no information. The citizens were information objects — providers of information but with no right to question, control, or correct it. Yet this was not simply a question of an absolute government that suppressed the citizens and prevented them from obtaining information. Even those who publicly criticized the government and argued for an open and free debate on all topics did not include the lower classes when it came to calling for unhampered access to information. The lower classes had an entirely different information culture in the streets of Copenhagen, with broadside ballads as one major information provider. Here information did not necessarily need to be true and reliable, as long as it evoked emotions and caused sensation (Skouvig, 2012). This was a distinctly different cultural understanding of information from that expressed by the central administration and proponents of civil rights.

2.3. Concluding Remarks

This chapter has presented a narrative of what appears to be the very antithesis of the present information society: scholars who submerged into a sea of vast information and took pleasure in filtering, sifting, and organizing it by means of new tools and techniques derived from the printing press. By the same token, Enlightenment savants did not conceive of information as a determinant of their time. Information was not an end in itself but always a means for something — a rhetorical weapon in the hands of the philosophers. The Habermasian conception of the public sphere has its roots in the Enlightenment and is more or less synonymous with free access to information and exchange of knowledge. This stands, however, in strong opposition to the policy of secrecy and control of information exercised by the absolutist state. This narrative can only provide a glimpse into past information cultures and strengthens Webster's call for a more profound genealogy. Such a genealogy should question assumptions about quantity, technology, and accessibility of information in order to challenge our present culture of information. A genealogy of information would reveal that information had specific purposes in specific settings. These purposes differed according to the situation for which it was created and thus information took on different forms in a learned community and in an absolutist state. And a genealogy would not imply a chronologically coherent or progressive history of how information conceptually developed from an early modern to a modern understanding of information. But it should give us the basic understanding of how information was culturally formed and perceived.

References

Black, A. (2006). Information history. In B. Cronin (Ed.). *Annual review of information science and technology* (Vol. 40, pp. 441–474). Medford, NJ: Information Today Inc.

Blair, A. (2010). *Too much to know: managing scholarly communication before the modern age*. New Haven, CT: Yale University Press.

Bonderup, G. (2012). *Enevælden 1660-1849*. Retrieved from http://danmarkshistorien.dk/leksikon-og-kilder/vis/materiale/enevaelden/?no_cache=1&cHash=a3c44f2aa27cb663f5e6d61d0ee2ccf2

Burke, P. (2000). *A social history of knowledge*. Cambridge: Polity Press.

Darnton, R. (2010). *Poetry and the police. Communication networks in eighteenth-century Paris*. Cambridge, MA: Harvard University Press.

de Vivo, F. (2007). *Information and communication in Venice. Rethinking early modern politics*. Oxford: Oxford University Press.

Engberg, J. (2009). Den standhaftige tinsoldat. En biografi om Frederik 6. Politiken, København.

Feldbæk, O. (2000). Vækst og reformer – dansk forvaltning 1720-1814. In L. Jespersen & E. Ladewig Pedtersen (Eds.), *Dansk forvaltningshistorie 1. Stat, forvaltning og samfund. Fra middelalderen til 1901*. Copenhagen: Jurist- og økonomforbundets forlag.

Finnemann, N. O. (2005). *Internettet i mediehistorisk perspektiv*. København: Samfundslitteratur.

Foucault, M. (1984a). Nietzsche, genealogy, history. In P. Rabinow (Ed.), *The Foucault reader* (pp. 76–101). New York, NY: Pantheon Books.

Foucault, M. (1984b). What is enlightenment? In P. Rabinow (Ed.), *The Foucault reader* (pp. 32–51). New York, NY: Pantheon Books.

Foucault, M. (2002). Of other spaces. In N. Mirzoeff (Ed.). *The visual cultural reader* (2nd ed., pp. 229–236). London: Routledge.

Giddens, A. (1985). The nation state and violence. *Vol. 2 of a contemporary critique of historical materialism*. Cambridge: Polity Press.

Habermas, J. (1990). Strukturwandel der Öffentlichkeit. Untersuchungen zu einer kategorie der bürgerlichen Gesellschaft. Suhrkamp Taschenbuch. Frankfurt am Main. Oprindelige udgave 1962.

Knorr Cetina (2007). Culture in global knowledge societies: Knowledge cultures and epistemic cultures. *Interdisciplinary Science Reviews*, *32*(4), 361–375.

Koselleck, R. (2013). Kritik und Krise. Eine Studie zur Pathogenese der bürgerlichen Welt. Suhrkamp Taschenbuch (1959/1973).

Krefting, W. E., Nøding, A., & Ringvej, M. (2014). *En pokkers skrivesyge. 1700-tallets dansk-norske tidsskrifter mellom sensur og ytringsfrihed*. Oslo: Spartacus Forlag.

Langen, U. (2012). *Det sorteste hjerte. Historien om et dybt fald I enevældens København*. København: Politikens Forlag.

Oldrup, T. (2007). For den historiske fuldstændigheds skyld. To københavnske avisers indhold i septemberdagene 1807. In P. Henningsen (Ed.), *København 1807. Belejring og bombardement* (pp. 158–181). Copenhagen: Jyllands-Postens Forlag.

Peters, J. D. (1988). Information: Notes towards a critical history. *The Journal of Communication Inquiry*, *12*(2), 9–23.

Peymann, E. (1807, September 1). Announcement from the commanding general to the garrison and all inhabitants in Copenhagen [Placard]. (37-122), The Collection of Pamphlets and Corporate Publications, The Royal Library, Copenhagen.

Postman, N. (1999). *Building a bridge to the 18th century. How the past can improve our future*. New York, NY: Vintage Books.

Raffnsøe, S., Gudmand-Høyer, M., & Thaning, M. (2008). *Foucault*. København: Samfundslitteratur.

Royal open Letter on rumors, dated July 26, 1812.

Rule, J. B. (1974). *Private lives and public surveillance*. New York, NY: Schocken Books.

Schou, J. H. (1808). Chronologisk Register over de Kongelige Forordninger og Aabne Breve, samt andre trykte Anordninger som fra Aar 1670 af ere udkomne, tilligemed eet nøigtigt Udtog af de endnu gieldende, for saa vidt samme i Almindelighed angaae Undersaatterne i Danmark og Norge, forsynet med et alfabetisk Register, 14. Copenhagen: Hofbogtrykker Christensen.

Skouvig, L. (2012). Produktivitet og moral: Almanakkens anmærkninger og skillingsviser omkring 1800. *Fund og Forskning*, *51*, 337–367.

Skouvig, L. (2015a). Efter indhentet efterretning: Et historisk rids af overvågning i enevældens København. *Nordisk Tidsskrift for Informationsvidenskab og Kulturformidling*, *3*(2–3), 7–17.

Skouvig, L. (2015b). Genres of war: Informing a city. In J. Andersen (Ed.), *Genre theory in information studies* (pp. 133–154) Studies in Information. Bingley: Emerald Group Publishing Limited.

Skouvig, L. (2017). Records and rumors. Surveillance and information in late absolutist Denmark (1770-1849). *Surveillance and Society*, *15*(2), 314–325.

Smith, P. H. (2011). What is a secret? Secrets and craft knowledge in early modern Europe. In E. Leong & A. Rankin (Eds.), *Secrets and knowledge in medicine and science 1500-1800* (pp. 47–69). Farnham: Ashgate Publishing Group.

Webster, F. (2006). The information society revisited. In L. Lievrouw & S. Livingstone (Eds.), *Handbook of new media. Social shaping and social consequences of ICTs* (Updated student edition). London: SAGE.

Webster, F. (2014). *Theories of the information society* (14th ed.). London: Routledge.

Weller, T. (2009). *The Victorians and information. A social and cultural history.* Saarbrücken: VDM Verlag.

Weller, T. (2011). Conclusion: information History in the Modern World. In T. Weller (ed.). *Information History in the Modern World.* Basingstoke: Palgrave Macmillan, p. 199–204.

Chapter 3

The (De-)Universalization of the United States: Inscribing Māori History in the Library of Congress Classification

Melissa Adler

Abstract

This chapter demonstrates how the University of Waikato in New Zealand adapted a global standard (the Library of Congress Classification) for local use by inscribing topics related to and about Māori history and people.

The findings are the result of using library catalogs and classifications as primary historical documents.

The University of Waikato's classification simultaneously uses and implicitly critiques a universal system written from a U.S. vantage point. It seems to acknowledge the benefits and necessities of using a globally recognized standard, as well as a need to inscribe local, anti-colonial perspectives into that system.

The research relies on historical documents, and some aspects related to purpose and attribution are difficult to ascertain.

The local adaptation of the Library of Congress Classification may serve as a model for other local adaptations.

This may bring new dimensions to thinking about colonialism and anticolonialism in knowledge organization systems. It contributes to ongoing conversations regarding indigenous knowledge organization practices.

The Organization of Knowledge: Caught Between Global Structures and Local Meaning
Studies in Information, Volume 12, 33–50

ISSN: 2055-5377/doi:10.1108/S2055-537720170000011009

Although scholars have examined Māori subject headings, research on local shelf classifications in New Zealand have not been objects of study in the context of global and local knowledge organization. This chapter brings an important classification to light.

Keywords: Anticolonialism; Universal knowledge organization systems; local classification; Māori; indigenous knowledge; New Zealand

Chapter Outline

> The twinning of power and legitimacy, one force obtaining in the world of direct domination, the other in the cultural sphere, is a characteristic of classical imperial hegemony. Where it differs in the American century is the quantum leap in the reach of cultural authority, thanks in large measure to the unprecedented growth in the apparatus for the diffusion and control of information.
>
> — Edward Said, *Culture and Imperialism*

> Research "through imperial eyes" describes an approach which assumes that Western ideas about the most fundamental things are the only ideas possible to hold, certainly the only rational ideas, and the only ideas which can make sense of the world, of reality, of social life and of human beings ... It is research which is imbued with an "attitude" and a "spirit" which assumes a certain ownership of the entire world, and which has established systems and forms of governance which embed that attitude in institutional practices.
>
> — Linda Tuhiwai Smith, *Decolonizing Methodologies*

"Universality" signals a variety of contexts and purposes in the field of knowledge organization, including the scope of topics and materials, the degree and type of access, and the anticipated users. For each of these conceptualizations of universality to be realized, there must also be a universalized center, or a set of assumptions around which structures and hierarchies are built. Plenty of scholars have identified the ways in which library classifications, including the Library of Congress Classification (LCC), the Dewey Decimal Classification, and Library of Congress Subject

Headings, have universalized patriarchy and whiteness, and mark deviations from assumed norms as "other" or exceptional. Many of these studies find that invisibility (usually of whiteness, maleness, heterosexuality, able-bodiedness, and Christianity) is the sign of an assumed universalized norm (Berman, 1971; Bowker & Star, 1999; Olson, 2002).[1] This chapter identifies the West, and more specifically, the United States, as the universalized and assumed norm around which subjects are organized in the LCC. It seems to be taken for granted that knowledge in libraries is organized from an American-centric perspective, but the processes of exclusion and marginalization upon which the universalization of the United States has been constructed remain relatively under-interrogated (Higgins, 2012; McKennon, 2006). I aim to open a space for asking about the consequences for history, identity, and the circulation of knowledge around the world, and for thinking about how the classificatory processes and practices that rationalize difference have been essential to framing and legitimizing colonialism and cultural imperialism. I am extending Mai's (2013) call to judge classifications from the social and cultural context in which they are produced, and I suggest that we must also consider the ethical tensions associated with the global applications of locally created systems. If all classifications that claim or aspire to be universal in coverage and reach are, in fact, particular, then we must interrogate the intended and unintended consequences of the globalization of knowledge organization systems. The history of knowledge organization systems that facilitate information retrieval must be understood as part of a much larger history of a struggle for power through definitional control with regard to political and social relations, territories, and subjects.

Situating the Library of Congress (LoC) as a United States federal agency enables an inquiry into how and for what purpose its classifications of knowledge about human subjects, political boundaries, and geographical locations have been designed. Such a frame foregrounds the ways in which its terminologies and hierarchical arrangements across the disciplines have fortified the state and enabled and sustained a "collective memory of imperialism" by organizing knowledge about racialized and colonized subjects. As Smith (1999) explains, indigenous knowledges have been "collected, classified and then represented in various ways back to the West, and then through the eyes of the West, back to those who have been colonized" (p. 1). The Library of Congress not only collects and arranges cultural, literary, and scientific documents and artifacts from all the disciplines and around the globe, but also sets the standards by which knowledge is

1 Far more studies could be named here. I regard Olson, Bowker and Star, and Berman to be foundational to this type of critique.

organized in libraries of all types around the world. Where the LoC classifies knowledge about nations, territories, and peoples, it must be viewed as a history of how the United States has researched, used, or organized them into U.S. history. And because the LoC's systems are used in catalogs around the globe, American-centric views are spread to libraries, rendering the catalog a colonized space. The LoC is a cultural archive of histories, texts, and artifacts from around the world, and it has created the technologies and systems that facilitate the retrieval and use of differentiated forms of knowledge in contexts other than those in which the artifacts were made. It is my hope that bringing attention to the question of cultural dominance by way of the library catalog will stir conversations about the consequences and complexities of knowledge organization in global and local contexts.

To illustrate the tension and conflict inherent in classifying the world's knowledge, I examine the ways in which works about and by Māori people have been classified by the LoC, in comparison and contrast with a local decolonizing shelf classification practices at the University of Waikato Library (Te Whare Pukapuka o Te Whare Wānanga o Waikato) in New Zealand. The LCC organizes Māori people as ethnographic subjects, or in other words, *objects* of study. I aim to show ways in which the LoC's classifications can be read as statements about political geography, knowledge, and power, and that those statements are at odds with and fail to represent the perspectives of Māori authors and communities. I will also demonstrate that local revisions to the classification at the University of Waikato assert the agency of Māori people in New Zealand's society and history. I read the locally revised classification as an anticolonial gesture that can serve as a model for revisions of knowledge organization systems in a variety of contexts. In its adoption and revision of LCC, it takes advantage of and participates in global conversations and resource sharing, but also inscribes and reorders topics to reflect and serve Māori communities, readers, and researchers.

3.1. The Catalog as Historical Record

The findings of this chapter are the result of using library catalogs and classifications as primary historical documents. Doing histories and textual analyses of classifications provides critical insights into the rhetorical and political aspects of knowledge organization. Not only do classifications facilitate access to knowledge, but they are also historical documents that contain knowledge about actors, agendas, and the ways in which catalogers, classifiers, and other knowledge workers have participated in wider political and social moments.

This study took shape like many archival projects do — by digging through documents, stumbling upon clues that lead to other clues, and stitching various pieces together from different sources. It is a logical continuation of my other work, in which I have examined ways in which classifications for sexuality, disability, and race have functioned as instruments of statecraft and the construction of a national history and identity (Adler, 2017). My concern here is to begin to figure out how this matters in a transnational context and to consider the consequences for information retrieval, as well as ways in which the spread of U.S.-centric ideology and norms by way of categories and systems that organize the facilitation of access to information affect history and identity formation around the globe.

I have found that one of the most effective starting points for historical/genealogical studies of classifications is to see how the theoretical works I am reading are cataloged. In this case, I began with the classification of Linda Tuhiwai Smith's *Decolonizing Methodologies* to test my theory about the export of cataloging techniques. Indeed, the classification of Smith's work is exactly the same in the the University of Waikato (Smith's place of employment) record as it is in the cataloging in publication data printed in the book and listed in the LoC catalog record. Comparing records in the Waikato catalog for other works, I found class assignments that didn't exist in the LCC, however, so I contacted librarians at Waikato, who generously shared locally revised sections of the classification. I continued to look closely at Classification Web, as well as the first, second, and third editions of the LCC to ascertain the moments in which Māori were written into certain spaces within the system. I compared the LoC's version with the revisions made at Waikato to understand how the latter more fully and accurately writes Māori history into the dominant system. I've brought my findings into dialogue with Smith and others who have identified ways in which U.S.-produced standards and cultural forms have affected local knowledges and peoples.

3.2. Organizing History

Using the example of the LCC, Bednarek (2007) has effectively argued that Western knowledge organization systems are constructed out of and reproduce colonial bias. One of the flaws that she identifies is the categorization of Māori topics within the D section, formerly defined as "General History and History of the Old World," and she suggests that relegating Māori subjects to "old world" history effectively erases Māori culture and peoples in the present. It isn't exactly clear how the classification distinguished general history from old world history in the D section; rather, the major organizing factor seems to be driven by the fact that the section is generally designed to classify everything other than North and South America. The current title for the D classification is world history and history of Europe, Asia, Africa, Australia, New Zealand, etc. I would argue that the placement in "History" is not the most significant problem, as, in fact, topics concerning nearly all regions of the world are also arranged in different sections of "History." Rather, the greater concern has to do with *how* various regional histories are organized in those sections in relation to and through the eyes of U.S. history. The last section of this chapter addresses the specifics of the classification, but first, I want to discuss the significance of narrating and organizing history from "universal" and local standpoints.

Smith's (1999) description of history is especially helpful for thinking about the function of the LCC. She identifies making history as a process of constructing a "story of the powerful and how they became powerful, and then how they use their power to keep them in positions in which they can continue to dominate others" (p. 34). She argues that this relationship between history and power results in and relies upon the production of "Others" and their exclusion. The ideas upon which Western history as a discipline and narrative are founded have everything to do with organizing knowledge. Smith presents the assumptions central to constructing dominant histories, beginning with the notion that history is a totalizing discourse — a master narrative that assumes that all knowledge can be organized into a coherent whole that tells the truth about the world. She regards classification as essential to determining which knowledges are selected and included in that history. That totality also includes the universalization of certain characteristics and values that all societies are thought to share and are deemed to be of "historical interest" (Smith, 1999, p. 30). History also requires and supports the idea that there is one totalizing chronology within which human progress can be charted, and that one of the functions of time is to define and exclude prehistorical events, individuals, and communities. Indeed, history is constructed around a wide array of binary categories that determine the inside and outside of history — the normal and abnormal, the self and Other — so as to determine and narrate a story of a domain. All of these assumptions and techniques for writing history provide support and legitimacy for dominance and imperialism.

Crucial to an analysis of colonized subjects is the acknowledgment that the LC classification orders the world in relation to how the United States imagines places and peoples to matter in relation to itself. Indeed, we might view the LCC as an instrument in the universalization of U.S. history. As the standard for classification has become universalized across the libraries that use it, so has the assumed universal center of the United States. As Butler (2000) writes, "The claim to universality always takes place in a given syntax, through a certain set of cultural conventions in a recognizable venue" (p. 35). But because there is no transnational consensus on how universality is constituted, the claim to universality must be translated across cultural and linguistic contexts, and "without translation, the only way the assertion of universality can cross a border is through a colonial and expansionist logic" (Butler, 2000, p. 35). Arguably, the LCC has been inadequately translated, as it cannot simply be a matter of translating from English to a local language. That kind of translation is in fact fueled by a logic of colonialism, as language becomes the instrument by which dominant values are exported. Rather, the concepts and the associations are provided as given, without a negotiation of meaning, and it has been up to the locales that adopt the scheme to adjust it for their purposes. Where local adaptations of LCC have been made, they tend to occur in very specific sections of the classification, and the overall structure and system remains according to the American design.

3.2.1. Exporting History

Said (1993) suggested that, although direct colonialism has mostly ended, imperialism lingers, especially in cultural spheres. He said that "structures of attitude and reference" (following Raymond Williams' "structures of feeling") appear in cultural productions of all sorts, from literature to anthropology, and that these structures extend across those works and are tied in some way to an ideology of empire. Across the board, Said (1993) found that within discourse communities and genres there was an underlying structuring attitude that "that one race deserves and has consistently earned the right to be considered the race whose main mission is to expand beyond its own domain" (p. 53). Conceptualizing the library catalog as a colonized space ruled and structured by U.S. terminologies and technologies, and occupied by an American imperial imagination, reveals that LCC is the kind of structuring mechanism to which Said refers, as it organizes all the disciplines into hierarchies and an inside and outside, that fit within the grand narrative of U.S. history and empire.

The LCC was designed in a particular moment in U.S. history, in 1898, just after the Library moved into its current Jefferson building. Librarian of Congress, John Russell Young imagined the LoC to be a critical agency in supporting U.S. national interests. He listed the LoC's collection development priorities as being in the areas of political, social, and religious movements that concern the United States (Young, 1898, p. 4). With regard to the purpose of the LoC, he wrote: "The Library of Congress must ultimately be the universal library of the Republic. To that end the most magnificent library edifice in the world has been erected and is destined to be, it is hoped, the home of America's literary and artistic genius, supplemented and strengthened by that of all lands and all time" (Young, 1897, p. 20). Classifiers were beginning to devise the LCC to organize this universal collection at this time.

To put it in a political context, work began on the present LCC around the time of the Spanish American War and while the United States was gaining power more globally. The United States was grappling with its struggle to unify after the American Civil War and the failures of Reconstruction, as well as imperialist projects in places like the Philippines, Hawai'i, and Puerto Rico. A close look at the LC classification reveals that it mirrors and organizes the ways in which the "race-making knowledge projects of anthropology, education, governance, and a propulsive investment in U.S. capitalism encoded hierarchies of human difference" in the colonial territories established at the end of the 19th century (Feldman, p. 104).

After World War II, American empire became a cultural and economic force, propelled and supported by information and communication technologies. The systems developed in the late 19th and early 20th centuries were much more widely disseminated through the information technologies developed in the postwar era. These processes also enabled the spread of information created by and about the United States around the world, including libraries. The U.S. government determined that postwar global democracy depended upon the diffusion of American cultural products and technologies (Richards, 1982). The pressures of World War II altered knowledge organization practice and priorities, and the LoC was central to the development of bibliographic control technologies to meet the demands of managing an onslaught of war and technical information (Stielow, 1990; Taube, 1951). Anglo-American cataloging and information technologies, many of which were developed by the Library of Congress, quickly became global in reach. For example, Machine Readable Cataloging (MARC) was developed at the LoC in the 1950s and 1960s and was instrumental in automating resource sharing and copy cataloging (Avram, 1975). Library of Congress Subject Headings are now used in libraries almost everywhere, either in English or in translation; according to the LoC's website, "Library of Congress Subject Headings (LCSH) is the only subject headings list accepted as the worldwide standard" (Library of Congress, Cataloging Distribution Service). The extent to which the LoC provides infrastructural technologies for libraries is astonishing, and there can be no question that people at the LoC were among those who "determined the intellectual agenda of and the perspectives of influential sectors of the government and academy" and participated in the expansion of American empire through ideological and cultural production (Said, 1993, p. 290). Classification is a critical component of the bibliographic control apparatus developed after

World War II, and looking back at postwar aspirations toward the unity of knowledge on a global scale call forth questions regarding the structuring of relationships across communities, nations, and disciplines to support U.S. efforts to spread American ideology and democracy.

It must also be noted that after World War II, the LoC organized and governed overseas missions to protect and obtain materials from war-damaged countries, and was involved in the huge project of installing American libraries and acquisitions offices in locations around the globe (Cole, 2005). The Library of Congress became a leader and partner in programs to bring modern librarianship to "underdeveloped" nations by bringing "experts" to train local staff and install Anglo-American technologies and standards. While these programs were conducted in the interest of preserving an international cultural record, saving books from war-torn countries, and enhancing democracy by getting reading materials into the hands of a vast public, there was also an underlying belief that the spread of American culture through books and other cultural products was a powerful and necessary instrument in winning the cold war (Barnhisel, 2015; Richards, 1982).

New Zealand was among the countries in which the Office of War Information established American-staffed libraries. Richards (1982) describes the American libraries in New Zealand opened in 1943 as exemplary of the projects that installed American ideology through library practice, and explains that the countries in the British Commonwealth were particularly important sites of postwar democratization and modernization because of the alliance between the United States and Great Britain. According to Richards, the activities of librarians as "practitioners of democratic American librarianship ... made some of them pioneers of local library development and turned their libraries into important conduits of the flow of U.S.-produced technical and professional information which was so greatly to affect the postwar world" (p. 329). Although the U.S. Information Library was closed in 1947 due to cuts in funding, American library standards and practices and education programs were established enough to be carried on. The installation of library classifications and technologies in libraries worldwide must be understood as a strategy in achieving and maintaining American dominance. Along with the spread of democracy and tools and techniques for facilitating access to information, the spread of American knowledge organization structures has been installed in libraries around the globe.

Where the United States has had a particular interest in narrating other nations' histories, we find that nationalist exceptionalisms and silences converge in cultural productions. Crucially, this analysis provides ample evidence in support of Said's (1984) assertion that a "disciplinary communications apparatus exists in the West" and that an "unconscious adherence to racial classification" means that certain narratives are barely in evidence (p. 30).

One could argue that the spread of Anglo-American librarianship to developing regions after World War II has much to do with this complicated mix of viewing other countries and peoples as inferior, a need to rescue, the advancement of democracy, and cultural imperialism. The exportation of the classification and other bibliographic standards into catalogs around the world has been one channel by which American ideology and culture has become dominant and deeply influential.

3.3. Classification and Decolonization of Māori Subjects

Permanent European (mostly British) settlements began in New Zealand around 1814. In 1840, tribal chiefs and the British Crown signed the Treaty of Waitangi, establishing New Zealand as a British colony, but also granting ownership of lands and protections to the Māori. Interpretations of the treaty were disputed, and the progressive terms of the treaty were largely ignored at the expense and marginalization of Māori people (Bardenheier, Wilkinson, & Dale, 2015; Harvey, 2015). In the 1970s, however, the Māori used the terms of the treaty to defend their rights, and in 1975 the Waitangi Tribunal began to formally investigate breaches of the treaty, leading to financial redress and cultural recognition in the 1990s. The country has officially recognized Māori language since 1987, when the Māori Language Act was passed (Harvey, 2015).

British settlers in New Zealand established an expansive library system. In fact, between 1872 and 1884 the Otago region had 116 public libraries and athenaeums for a population of just over 135,000 people (Harvey, 2015, p. 183). By the beginning of World War II, New Zealand had around 450 public libraries (Richards, 1982, p. 334).[2] As of 2015, New Zealand has 317 public libraries (Public Libraries of New Zealand, 2015). New Zealand has a strong, well-established library system, but it is mostly based on standards, practices, and principles of Anglo-American librarianship.

Māori knowledge organization practices have featured relatively prominently in the emerging field of inquiry into indigenous knowledge organization.[3] This is due, at least in part, to the fact that New Zealand has worked toward accounting for injustices, pursuing reparative measures, and including Māori people and culture in New Zealand history. Māori Subject

2 Public libraries in New Zealand at that time were mostly subscription libraries that generated revenue by charging fees (Harvey, 2015; Richards, 1982).

3 For example, the 2015 special issue of *Cataloging and Classification Quarterly* on Indigenous KO, edited by Doyle and Metoyer, included three articles on Māori KO, and the introduction was written by Linda Tuhiwai Smith.

Headings were among the earlier projects aimed toward naming, defining, and arranging subjects according to the language and epistemological frameworks of indigenous communities. Work on the Māori subject list began in 2005, and a year later the Nga Upoko Tukutuku online thesaurus was launched (Lilley, 2015; Te Rōpū Whakahau, 2016). The headings are included in the New Zealand National Bibliographic Database as authorized headings.[4] With regard to shelf classifications, most academic libraries in New Zealand libraries use LCC and the Dewey Decimal System, but there are some instances of local revisions or augmentations of those systems. For example, Bardenheier et al. (2015) have described the system adopted at the Sylvia Ashton-Warner Library, which collects and organizes youth Māori language material for the Faculty of Education at the University of Auckland. That library uses the Dewey Decimal Classification for consistency and resource sharing with New Zealand school libraries, but it has also applied the Ngā Kete Kōrero Framework to categorize Māori language reading levels (p. 499). That framework is based on Māori views about knowledge acquisition, and uses basket weaving as a metaphor to represent literacy development.

The revisions to the LCC at the University of Waikato serves as another example that demonstrates how knowledge created by and about Māori communities has been locally inscribed into a global standard. The LCC is used in the university at which Linda Tuhiwai Smith teaches and conducts research, thus illustrating one case in which collecting, cataloging, and classifying works from and about New Zealand according to Western standards are then exported and repeated in the location of origin. Smith's book is not classed with books about New Zealand or Māori people, but rather, is located in Anthropology at GN380, defined as "Ethnographies of special categories of people." The scope note provided in Classification Web for this location provides instructions: "Class here ethnographies that discuss collectively the nonliterate people of the world whose particular way of life is being shattered by Western encroachments."

The University of Waikato was established in 1964 in Hamilton, New Zealand. The Centre for Māori Studies and Research was formally organized in the School of Social Sciences in 1972. The university has focused on incorporating Māori studies in the university curriculum, student life, and the library. It is not entirely certain that the classification was developed at Waikato, as the historical record does not indicate where the classification was created, nor the individuals who worked on it. A similar system is used at the University of Canterbury in New Zealand. It seems

4 Additionally, the first open-source integrated library system, Koha, is a New Zealand creation (Eby, 2007).

UNIVERSITY OF WAIKATO LIBRARY

NEW ZEALAND AND PACIFIC COLLECTION
HISTORY CLASSIFICATION

Figure 3.1: Section title for DU400-DU445.

appropriate to suggest that it was, in fact, developed at the University of Waikato because the first page of the local section is headed with the name of the library (Figure 3.1). Additionally, discrepancies between the Canterbury and Waikato catalogs indicate that Waikato more closely adheres to the adaptation. According to Tony Millett, retired deputy librarian, it is likely that Jim Benzoni, who began working at the University of Waikato Library in early 1967, would have played a major part in adapting the classification, possibly with some assistance from Frances Alexander. Additionally, it may have been inspired in part by O'Reilly's 1964 expansion of the Dewey Decimal New Zealand history classification numbers, which is currently held by the Victoria University of Wellington Library.[5]

The shelf classification used at the University of Waikato does indicate a decolonizing technique that foregrounds Māori history and peoples. Below I compare and contrast the Library of Congress's version with the version used in New Zealand.

Smith (1999) points to the ways in which Western research methods treat communities and individuals as objects of study, suggesting that this is one way to maintain the divide between the self and Other and maintain a sense of superiority. We find evidence of this kind of treatment in the LCC, as it has consistently treated Māori people as ethnographic subjects. The present classification, available via Classification Web, provides a detailed arrangement of Māori subjects within the range DU422.8-DU424, all within the broader class "Ethnography" in the history of New Zealand. The classification of the D section, formerly titled "Universal and Old World History," was first printed in 1916. In the preface to this first edition, Alfred F. W. Schmidt, the "Classifier in Charge," indicates that the classification reflects the collection of the LoC and that certain sections lack detail, due to the limited material in the collection. However, a close look at the catalog using the keyword "Maori" and then examining the records for the Library of Congress Control number, which includes the date of cataloging, reveals that dozens of works were in the LoC collection by 1916. Although there

5 This information was obtained via email correspondence with Jenny Hooker, librarian at the University of Waikato.

was ample material in the collection, Schmidt and his staff did not view Māori to be worth including as a subject in history. The Māori people were not named in the first edition of the D classification, but there was a space at DU423 for the more general "Ethnography and races" under New Zealand. In 1959 the caption for the class was changed to "Ethnography. Races. Māoris" without any further subdivision. The LCC first included Māori as a distinct subclass of Ethnography in 1989, at which time the LoC created a detailed arrangement of topics within Māori. These topics are primarily organized from the point of view of ethnographers and include five broad subclasses: "Biography" (422.8-82.A-Z), "General works" (423.A1), "Study and teaching" (423.A15), "Special topics, A-Z" (423.A2-Z), and "Individual tribes and cultures" (424.A-Z). The Special topics, A-Z section includes around two dozen topics, such as "Boats. Canoes," "Funeral rites and ceremonies," "Hunting," "Implements," and "Politics and government." The 1989 arrangement is essentially the same in 2016, but with twice as many categories. Similarly, whereas the 1989 subdivided "Individual tribes and cultures" into three tribes, the present edition names over 30.

Ethnographic works on New Zealand are located in the Anthropology section (GN), and some works on Māori are classed in this location at the LoC. The class GN667.N9 was first established in 1954, under "Races and ethnic groups, By country" in the third edition of the G section, which was then broadly titled "Geography, Anthropology, Folk-lore, Manners and customs, Recreation." At that time, Māori were not mentioned in the GN section, nor were they yet named in the DU section. This fuzziness might help to explain why the LoC placed some early works in the GN section. Māori were not included at all in the DU section until five years later, when the class was listed under "Ethnography," and it did not become highly subdivided into specific topics related to ethnography for another 30 years. Today, there is a note in Classification Web directing catalogers to class ethnographic works on Māori in DU — not in GN with other New Zealand ethnographies.

It seems that the necessary interpretation of the designation of Māori subjects as ethnographies in the DU section is that classifiers understood New Zealand to include the white and mostly British population, and that the indigenous population was relevant to history only in terms of ethnography. The University of Waikato has removed references to ethnography in their revised and local version of the LCC, so Māori subjects are considered part of New Zealand history, without being treated as objects of ethnographic study. Additionally, at Waikato, works that are explicitly and specifically ethnographies of Māori communities are located in the Anthropology section (GN667) along with other ethnographies of New Zealand (Figure 3.2).

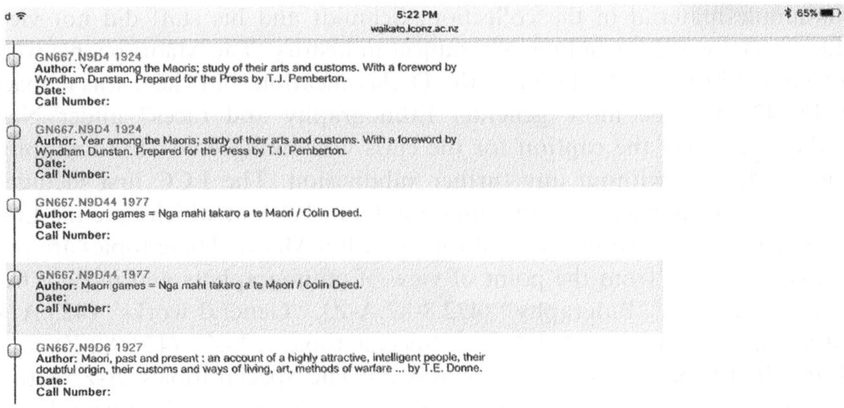

Figure 3.2: Screenshot from a catalog search of the Waikato catalog showing a selection of the books on Māori ethnographies cataloged at GN667.N9.

This brief list demonstrates an important difference between the Library of Congress's practices and Waikato's. Each of these titles seems to have been determined to be ethnographic works, and Waikato has placed them all within the Anthropology section. The LoC, however, has placed these in different locations. The first title listed, *Year Among the Māoris: Study of Their Arts and Customs*, is in GN667.NP, but the other two — *Māori Games* and *Māori, Past and Present: An Account of a Highly Attractive, Intelligent People, Their Doubtful Origin, Their Customs and Ways of Living, Art, Methods of Warfare* — have both been placed in DU423 at the LoC, where Māori topics are classed as ethnographies in history.

The local classification used at Waikato inserts Māori history at DU412-445. In sharp contrast to Classification Web, it specifically states that Anthropological studies should be placed in GN667.5, and overall, the divisions are much more specific to Māori history.

412 Māoris. General works (Anthropological studies GN667.5)

414 Miscellaneous including general reminiscences without specific subject content

History, general

415 Addresses, essays, lectures

416 General special

417 Comprehensive

At DU418-434, Waikato's classification is subdivided by time periods and wars. This is roughly the same area of the LCC that divides New

Zealand history by time period.

418	To 1840
.1	N.Z. Company
.2	Treaty of Waitangi, etc.
420	1840–1900
421	Māori Wars. General.
.1	First 1845-48
.2	Second 1860-7
425	1900-

The hierarchy continues beyond the year 2000. In contrast to the Waikato version, the Treaty of Waitangi only appears in the LCC in the discipline of law, at KUQ353. Its absence in the history of New Zealand is striking because it is a foundational and controversial document in New Zealand's history. The placement of the Treaty in the local classification is a critical decolonizing act that speaks against the erasure of Māori history.

In the space where LC classed Māori as ethnographic subjects (DU422.5-424) and Waikato removed the "ethnography" label, Waikato has retained some of the arrangements within. For example, the Special topics, A-Z at DU423.A2-Z are present. Hand-written classes have also been added, providing subdivisions such as "Prejudices" at DU423.P74, for example. Indeed, the hierarchies look very similar to the original LC classification.

	Māoris
	Biography
	Including portraits
.8	Collective
.82	Individual, A-Z
423.A1A-Z	General
.A15	Study and teaching
.A2-Z	Special topics, A-Z

By claiming certain spaces for local history within a universal system, the Waikato classification is not simply a decolonizing gesture, but an assertion of rights and access to knowledge and an act that facilitates seeking and discovery of knowledge in ways that more accurately reflect Māori perspectives and interests. While it does not reorganize knowledge according to Māori principles in the way that other local systems have done, it does very effectively install subjects and history where they were omitted.

3.4. Conclusion

Although the classifications that organize materials on shelves are mostly hidden from library users' views, the colonial biases they carry into libraries around the globe must be called into question. The classification is built not only upon the universality of heteropatriarchy and whiteness, but also on the universality of the United States. This means that nations, states, and territories are set apart in opposition into an assumed U.S. norm and their histories and subjects are organized in relation to U.S. interests, rather than their own. The case of the University of Waikato classification is particularly compelling for the way it simultaneously uses and implicitly critiques a universal system written from a U.S. vantage point. It seems to acknowledge the benefits and necessities of using a globally recognized standard, as well as a need to inscribe local, anticolonial perspectives into that system.

Acknowledgments

I owe many thanks to Jenny Hooker and the staff and the University of Waikato Library for supplying me with the classification and very helpful information about its origins.

References

Adler, M. (2017). *Cruising the library: Perversities in the organization of knowledge.* Bronx, NY: Fordham University Press.

Avram, H. (1975). *MARC: Its history and implications.* Washington, DC: Library of Congress.

Bardenheier, P., Wilkinson, E. H., & Dale, H. (Te Rarawa, Te Aupōuri). (2015). Kite Tika te Hanga, Ka Pakari te Kete: With the right structure we weave a strong basket. *Cataloging & Classification Quarterly*, *53*(5–6), 496–519.

Barnhisel, G. (2015). *Cold war modernists: Art, literature, & American cultural diplomacy.* New York, NY: Columbia University Press.

Bednarek, A. (2007). The colonial bias: Library classification in Aotearoa New Zealand. *Library and Information Management Journal*, *50*(3), 233–245.

Berman, S. (1971). *Prejudices and antipathies: A tract on the LC Subject Heads concerning people*. Metuchen, NJ: Scarecrow.

Bowker, G. C., & Star, S. L. (1999). *Sorting things out: Classification and its consequences*. Cambridge, MA: MIT.

Butler, J. (2000). Restaging the universal: Hegemony and the limits of formalism. In J. Butler, E. Laclau, & S. Žižek (Eds.), *Contingency, hegemony, universality: Contemporary dialogues on the left* (pp. 11–43). London: Verso.

Cole, J. Y. (2005). The Library of Congress becomes a world library, 1815-2005. *Libraries & Culture*, *40*(3), 385–398.

Doyle, A. M., & Metoyer, C. A. (2015). Indigenous knowledge organization [Special issue]. *Cataloging & Classification Quarterly*, *53*(5–6).

Eby, R. (2007). *Open-source server applications*. Library Technology Reports, (May–June), 48–53.

Feldman, K. P. (2015). *A shadow over Palestine: The imperial life of race in America*. Minneapolis, MN: University of Minnesota Press.

Harvey, R. (2015). Australasia. In P. S. Richards, W. A. Wiegand, & M. Dalbello (Eds.), *The history of modern librarianship: Constructing the heritage of western cultures* (pp. 179–204). Santa Barbara, CA: Libraries Unlimited.

Higgins, C. (2012). Library of congress classification: Teddy Roosevelt's world in numbers? *Cataloging & Classification Quarterly*, *50*(4), 249–262.

Library of Congress. (1916). *Classification, class D, universal and old world history*. Washington, DC: Government Printing Office.

Library of Congress. (1954). *Classification, class G, geography, anthropology, folklore, manners and customs, recreation* (3rd ed.). Washington, DC: Government Printing Office.

Library of Congress. (1959). *Classification, history: Class D, general and old world* (2nd ed.). Washington, DC: Government Printing Office.

Library of Congress, Cataloging Distribution Service. Library of congress subject headings. Retrieved from http://www.loc.gov/cds/products/product.php?productID=214. Accessed on July 8, 2015.

Lilley, S. C. (2015). Ka Pō, Ka Ao, Ka Awatea: The interface between epistemology and Māori subject headings. *Cataloging & Classification Quarterly*, *53*(5–6), 479–495.

Mai, J. E. (2013). Ethics, values and morality in contemporary library classifications. *Knowledge Organization*, *40*(3), 242–253.

McKennon, E. (2006). Importing hegemony: Library information systems and U.S. hegemony in Canada and Latin America. *Radical History Review*, *95*, 45–69.

Olson, H. A. (2002). *The power to name: Locating the limits of subject representation in libraries*. Dordrecht: Kluwer.

O'Reilly, R. N. (1964). *The decimal subdivision of the New Zealand class: The third stage of the N.Z. Library Association revised (1963) classification schedules, designed for regional collections*. Christchurch: Canterbury Public Library.

Public Libraries of New Zealand. (2015). *New Zealand Public Library Statistics 2014/2015*. Wellington: Public Libraries of New Zealand.

Richards, P. S. (1982). Information for the allies: Office of war information libraries in Australia, New Zealand, and South Africa. *The Library Quarterly*, *52*(4), 325–347.

Said, E. (1984). Permission to narrate. *Journal of Palestine Studies*, *13*(3), 27–48.

Said, E. (1993). *Culture and imperialism*. New York, NY: Vintage.

Smith, L. T. (1999). *Decolonizing methodologies: Research and indigenous peoples*. New York, NY: Zed Books.

Stielow, F. (1990). Librarian warriors and rapprochement: Carl Milam, Archibald MacLeish, and World War II. *Libraries & Culture*, *25*(4), 513–533.

Taube, M. (1951). Functional approach to bibliographic organization: A critique and a proposal. In Bibliographic Organization: Papers presented before the 15th annual conference of the graduate library school (pp. 57–71). Chicago, IL: University of Chicago Press.

Te Rōpū Whakahau. (2016). Ngā Ūpoko Tukutuku/Maori Subject Headings. Retrieved from https://www.trw.org.nz/current-projects/nga-upoko-tukutuku-maori-subject-headings/. Accessed on June 10, 2016.

Young, J. R. (1897). *Report of the librarian of congress*. Washington, DC: Government Printing Office.

Young, J. R. (1898). *Report of the Librarian of Congress* for the Fiscal Year ended June 30, 1898. Washington, DC: Government Printing Office.

Chapter 4

Reader-Interest Classifications: Local Classifications or Global Industry Interest?

Daniel Martínez-Ávila

Abstract

This chapter reviews the historical tension between global and local interests in library classifications. More specifically, this chapter presents the concept and characteristics of the reader-interest classifications as they were reported in the literature of the past century, including its alleged advantages and detected shortcomings, in order to discuss their presence and consequences in current cases of reader-interest classifications such as BISAC. Following an implicit post-structuralist approach, issues such as the role of standardization and centralization in these projects, the focus and philosophy underlying the construction of these classifications, and the underlying global interests of the book industry are analyzed in order to determine the social consequences and viability of these local classifications. It is concluded that libraries that consider adopting a reader-interesting classification must really think of the interest of the users (in plural) and not only of the global book industry that dominates the development of the standards.

Keywords: Reader-interest classifications; BISAC; standardization; centralization

The Organization of Knowledge: Caught Between Global Structures and Local Meaning
Studies in Information, Volume 12, 51–69

ISSN: 2055-5377/doi:10.1108/S2055-537720170000011010

Chapter Outline

4.1. Introduction

The tension between rigid standardization and local readers' interests has been affecting libraries since at least the last century. In the late 1930s, Detroit Public Library, in the United States, started a local project of a "reader-interest classification" as an alternative to standardized library classifications such as the Dewey Decimal Classification (DDC) (Rutzen, 1952). Reader-interest classifications were claimed in the literature to represent the "local interest" of readers. For the following decades, many "reader-interest classifications" flourished in the United States, the United Kingdom, and other countries around the world having some characteristics in common (Martínez-Ávila & San Segundo, 2013). Since the 2000s, several libraries in the United States have continued these practices adopting bookstore-like schemes such as BISAC, maintained by the Book Industry Study Group (BISG), for the organization of their collections. These recent classifications have been opposed to the "global imposition" of DDC and the "global interest" of Online Computer Library Center (OCLC) and are considered new cases of reader-interest classifications (Martínez-Ávila, Segundo, & Olson, 2014). However, can we really say that these cases of reader-interest classification are considering the interests of the totality of users? Are at least multiple and diverse views being considered in these local alternatives? The hypothesis of this chapter is that the approach followed in the construction of reader-interest classification, arguably based on user-based and cognitive studies, has serious problems to effectively justify the inclusion of the totality of users, and thus to justify the validity of its inductive reasoning. Following an implicit post-structuralist approach (in which it is accepted not only the multiplicity of realities but also the ethical need to consider the "othered"), issues such as the role of standardization and centralization in these projects, the focus and philosophy underlying the construction of these classifications, and the underlying global interests of the book industry are analyzed in order to determine the social consequences and viability of these local classifications.

4.2. Reader-Interest classification Definition and Characteristics

According to Satija's "dictionary of knowledge organization" (2004, p. 182), a reader-interest classification is:

> A classification designed to serve the immediate needs of the targeted users. Such systems may violate the filiatory sequence to bring together disparaged subjects needed by a user group. Indeed these are useful in mission oriented or multidisciplinary subjects. In a commerce college, e.g., it may be more pragmatic to place commercial law with commerce at 380. It is true to say that reader's interest classification adopted so far are not satisfactory in the long run and sometimes correspond to ephemeral vogues. It reflects a middle level of ambition in knowledge organisation. It is a compromise between ad hoc classification and rigorously scientific classification.

This type of classification was usually presented as an alternative to the model of traditional library classifications such as the DDC or the UDC, which were alleged to be unhelpful for the reader. One of the main aims of this new type of arrangement was to enhance the collection's usefulness for the end-user — or to at least enhance its perceived usefulness. This perception was influenced by features which were not always related to the classification system — such as display and presentation of the scheme and also of the collection and the shelves. This being the case, most reader-interest classification projects were indivisibly linked to many other aspects related to guiding and signage.

In practice, most reader-interest classification features varied according to the specific characteristics of each case. As Sapiie pointed out: "The physical arrangement which puts this principle into action is determined by local conditions, such as community needs and interests, physical size and layout of the building, and the size and nature of the collection" (Sapiie, 1995, p. 147). Despite such specific characteristics, Sapiie and authors such as Hubbard (1972) also recognized that there were some common patterns among all these cases. The arrangement of categories within the system depended on the characteristics of each case and was usually random, alphabetical by author or by Dewey. In larger categories order was usually by Dewey, while in smaller categories it was more likely to be random. Accordingly, alphabetical arrangement was the most appropriate order for the first level of categories, while random order was most appropriate for a second level (with the possibility of applying some other more specific method such as by year or by quality).

4.3. The Design of Reader-Interest Classifications

Most reader-interest classifications usually acknowledged the influence of previous cases, with some even citing the first examples at the Detroit Public Library. Yet it is also true that most reader-interest classifications were designed *ad hoc* according to the specific characteristics of the library intending to implement it. Although many libraries claimed to have been influenced by a reader-interest classification project, there do not seem to be any reported cases where libraries strictly adopted a previous reader-interest classification without modifying it to some extent. On the other hand, many libraries also looked to bookstore practices to avoid the alleged problems and influence that Dewey had over their categorization systems, including aspects such as the development and assignation of categories. Cambridgeshire libraries, in the United Kingdom, for example, developed their headings by examining catalogs and categories that were used in bookshops (Chandler, 1982). In every case, one of the most important factors in designing the scheme was how to adapt to characteristics of the local library and to thus meet the needs of the reader. Without this adaption to local needs, reader-interest classification schemes were said to run the risk of being just as inadequate for library readers as the standardized Dewey system that these schemes were often intended to replace. In words of Sapiie (1995, p. 146):

> They [libraries adopting reader-interest classifications] relied on their own good judgment and experience in defining the categories, as the local needs do not necessarily match the published lists of fiction

> categories and genres or the most popular categories and genres of fiction often included in published surveys. It is recognized that intimate knowledge of the library community is required to define the categories and then to fill both fiction and non-fiction categories with books which match the readers' ideas of what belongs there.

While each library was unique, most reader-interest classifications were also based on similar principles. As Ainley and Totterdell stated, "the schemes in fact share significant similarities despite the disparity in community makeup and initial approach" (Ainley & Totterdell, 1982, p. 119). Although such similarities were never strict, they were significant and a reflection of the continuous struggle between adoption and adaptation of previous cases in the reader-interest classification projects. On the other hand, readjustments to circumstances and to deviations from standard reader-behavior were considered a fundamental part of the (re)design process of reader-interest classification categories. Although such continuous readjustments led to a high level of flexibility within the scheme, they also meant there was a lack of inter-library standardization. Since there was no organization which was generally responsible for an "authorized" version of a reader-interest classification, more readjustments meant a greater deviation from other reader-interest classifications.

4.4. Reader-Interest Classifications Alleged Advantages

Because reader-interest classifications were considered part of the user-center movement in libraries, the general arguments of user-center practices and arguments in Library & Information Science were also well accepted and adapted to the specific case of classifications. As such, many alleged advantages and aspects of the reader-interest classification were presented as being improvements upon the apparent shortcomings of traditional classifications. Reader-interest classifications were designed from experience; for its supporters, reader-interest classifications improved user experience by addressing the inadequacies of traditional schemes and their system-oriented perspective. Some of these aspects and alleged advantages will be discussed in more depth later.

4.4.1. Meeting the User's Need by Gathering Materials of Interest: The Distributed Relatives

At the time that reader-interest classifications were being considered, libraries carried out surveys that found that the majority of users were not satisfied with public library service. One reason for this, according to the surveys, was that books were displayed in Dewey sequence, a 19th century conception of Librarianship that divided knowledge into academic disciplines instead of topics of interest. In the literature on reader-interest classifications, this problem received the name of distributed relatives; for most advocates of reader-interest classification — see for instance (Sawbridge & Favret, 1982) — the main cause of this problem lay in the very nature of the DDC itself. In a well-known quote of the reader-interest classification literature, Betts (1982, p. 63) explained distributed relatives as follows:

> Traditional classification schemes, in creating a logical set of relationships between "subjects," fail to take account of the (changing) interests which lead people to approach those subjects. Interests cross logical boundaries (as do books themselves at times) with the consequence that books which readers would wish to access by interest are often widely and inconveniently separated on the shelves and in some instances one or other sequence may never be found. Conversely, books appear together on the shelves which have no relationship other than a formal academic one, to the benefit of no one in particular. The positive corollary of all this is that books should be grouped to reflect the actual or potential interest relationship between them, even if this means fragmenting the traditional classification sequence.

Other authors have also pointed out the problem of distributed relatives: "Although subjects have a tendency to be treated in classification theory as thought they were natural, objective (i.e., 'real') categories, in fact they are arbitrary assemblages dependent upon time and place: one culture's religion (Dewey 200s) can easily be another's folklore (Dewey 390s). The categorization which necessarily precedes classification is therefore the product of a specific cultural and intellectual milieu" (Donovan, 1991, p. 28). Yet, according to Donovan, there might be more cultural divergences and disagreements among some disciplines when gathered by Dewey than when they remain as scattered subjects (as for example the different traditions in Anthropology between the United States, Great Britain, and Germany). In addition, Donovan also claimed that arrangement by discipline generates at least two more problems: first, it introduces intellectual assumptions that are not always valid outside their original contexts; and second, it creates confusion since most works are not pure in their disciplinary contents, but rather mix or contrast disparate fields of study. In conclusion, Donovan claimed that library classification systems based on academic disciplines cannot easily and neatly accommodate heterogeneous (interdisciplinary) materials in a way that maximizes utility for the patrons — that is, the way that the system obliges users to perceive subjects and the collection is psychologically uncomfortable.

4.4.2. Flexibility and Shelf Arrangement of Reader-Interest Classifications

Ainley and Totterdell considered that rejection of traditional classification systems extended to the "standardisation of variations in Dewey Decimal Classification at local government reorganisation" and that there was "an increasing awareness by all that this traditional classification system is inflexible and unhelpful to readers" (Ainley & Totterdell, 1982, p. 120). In many cases where Dewey was used as the basis for developing categories remainders or influences from the old system were seen as a burden upon results. McCarthy (1982, p. 93) called this the "British Museum syndrome" (McCarthy, 1982, p. 93) — the idea that all Dewey numbers should be represented in the categories — and qualified it as something to be avoided during the reader-interest classification development process. Betts similarly pointed out "the dangers in adopting a Dewey basis for allocating books to particular categories" (Betts, 1982, p. 65); these included misplacement of the book according to reader-interest because of the primary Dewey topic, and the gathering of divergent interest-oriented books in the sub-division of Dewey topics. In addition, Betts found that the "logical" nature of Dewey ran counter to the "interest" nature of the reader-interest classification categories and claimed that "some freedom of action is needed for classifiers and categorisers."

Within a traditional Dewey arrangement, ordered according to academic disciplines, it was not possible to abandon strict shelf order within both fiction and nonfiction categories. Leaving this order, however, was a good option for libraries who found that Dewey scattered materials of interest and was not flexible enough to meet the needs of the user. As Hubbard pointed out, loose order provided reader-interest classification with one of its greatest claims to flexibility: "Non-sequential verbal categories can be put wherever they're most useful in a particular library at a particular time. 'Spanish' can be shelved next to other foreign language books, or next to 'Spanish Lit,' or next to 'Spain'" (Hubbard, 1972, p. 26). Sawbridge and Favret were among other authors who considered flexibility to be one of the main advantages of reader-interest classifications: "One of the advantages of such a scheme is that if a book fails to find readers in one category it may be moved to another with little difficulty" (Sawbridge & Favret, 1982, p. 386). A similar view was held by Lyn Donbroski, who pointed out that the lack of flexibility and convenience in using Dewey had been key factors in the adoption of reader-interest classifications in the United Kingdom (Donbroski, 1982, p. 78).

4.4.3. More Adequate Browsing and Increase of Circulation in Reader-Interest Classifications

One of the most important claims of reader-interest classifications was that they were more adequate for browsing in libraries than traditional classifications. This claim was usually accompanied by at least two other major assumptions: first, that most library users are browsers (or nonpurposive readers as they are sometimes called in the reader-interest classification literature); and second, that traditional library classifications such as Dewey are not adequate for browsing (as users do not understand notation). Related to this advantage, another alleged advantage of reader-interest classifications was that it increased circulation. This aspect was also linked in the literature to the use of reader-interest classifications as a method for stock control (see Martínez-Ávila & Satija, 2015).

4.5. Reader-Interest Classifications Shortcomings

Some of the reader-interest classification shortcomings include the "other" distributed relatives, quality and nature of categories, reclassification of stock, and the tension between standardization and local adaptations. These reader-interest classification shortcomings will be discussed in more depth later.

4.5.1. The "Other" Distributed Relatives and the Singular Reader

The problem of distributed relatives in traditional classifications such as Dewey was claimed to have been resolved by rearranging facets according to the reader's interest. This solution, however, was also object of further criticism and discussion; as pointed out by several authors, every facet order inevitably benefits some group of users and yet, in creating new distributed relatives, it has an adverse effect on the other groups. For Payne (1983), the main cause of the distributed relatives problem was not just to be found in the nature of the DDC (as maintained by Sawbridge & Favret, 1982, p. 385): "this is a half truth, for it is not a particular problem of Dewey. As long as authors persist in writing books on complex subjects any linear arrangement of those books will result in distributed relatives: create broader classes and re-christen them 'categories' or whatever, if you will, the problem will not go away" (p. 29). According to Payne, all the reader-interest classification examples given by Sawbridge and Favret, concerning topics of interest that were separated by Dewey and gathered by the reader-interest classifications, created new and potential distributed relatives and these could only be understood or resolved in terms of the linear nature of library classifications (and not in terms of the academic-like discipline which was the basis of Dewey — as suggested by reader-interest classification proponents). In short, every new arrangement scatters relatives and reader-interest classification does not seem to mean the end of this problem. In the words of Payne: "It is all a case of choosing which relatives to distribute, bearing in mind that in the out-and-out categorized library, with the abandonment of a closely classified and alphabetically indexed arrangement, any chance of ever gathering them up again is lost."

In general, the problem of the new distributed relatives in reader-interest classifications was well known and even accepted among advocates of reader-interest classification. As a solution, they proposed relegating the interests of a certain group of users (purposive users) for the benefit of "the reader" (meaning the browser). The structure of the system was replaced by a new and entirely reader-oriented structure: the reader-interest classification. This was not a solution that would allow for the integration of all kind of users. Instead, what was created was a brand new scheme that followed the same logic, the same principles, and which caused the same inconveniences for certain groups of users as the old scheme. The only difference here was that the group of users that was placed at the center of the scheme — which benefited most from the system — was the group generally accepted as "the reader."

However, as Hope Olson has pointed out, a mere replacement of the rejected system with a new system, that targets a new group of users but follows the same logical principles, might not be the best solution: "All systems have a limit that excludes and/or marginalizes something or someone. Replacing the old system with a new one may change the limits, but it does not eliminate them. A more useful approach is to make the limit permeable to allow the voices of those previously excluded or marginalized to speak within the system" (Olson, 1997, p. 62). What is more, simply replacing the limits of reader-interest classifications could hardly be considered as vital to improving marginalized groups' access to information. In this vein, only Sue Bridgwater of De Beauvoir Junior School Library showed any critical sensitivity toward marginalized and minority groups within the discourse of reader-interest classification (Bridgwater, 1990, p. 53):

> Both school staff and those of us involved in reorganising, tidying up and operating the library felt strongly that Dewey has several disadvantages today. [...] Dewey is profoundly racist by modern standards, with Western European cultures placed first in every schedule and given more room for detailed classification than the rest of the world. This is a particular disadvantage in an inner city East End school whose community is representative of many ethnic groups and where

the languages spoken by pupils include Bengali, Greek, Turkish, Spanish, and a couple of African languages.

However, such concerns did not appear to have been taken into account when the system was finally implemented. There was thus a new and privileged group of users in reader-interest classifications — the group that most benefited from facet order in the system. Although this group was formed by democratic and majority decision, minority interests were not taken into account in this process.

Within the discourse of reader-interest classification, the targeted audience of the system is commonly referred to as "the reader," "the user," "the customer," or "the public." Use of the singular forms of these terms, including the explicit *reader*-interest classification name and the *user*-centered context, often denotes that only the interests of one single, homogeneous, and exclusive group are intended to be addressed by the system. As Olson and Schlegl (2001, p. 78) pointed out "the danger of speaking of 'the user' is that that phrase suggests a particular user, a 'majority reader', rather than the heterogeneous groups of users that vary from one context to another." This use of a singular public or a singular reader in the reader-interest classification discourse might be partly regarded as the legacy of Cutter's "Rules for a Dictionary Catalog" on library standards. It was Olson who noticed that the use of a singular public in Cutter's discourses indicates a type of exclusivity — a vision of a singular public with one convenience and a singular way of seeking information (Olson, 1997, p. 56): "a universality is present in Cutter's view, but it is the singular public who defines it. If a universality is present then it implies that there is not diversity. What diversity is excluded if the universality represents the community? [...] The result, then, of a community in singular is that it is not totally inclusive. It excludes those who are somehow different."

Still, there were some other authors, such as author Roy Payne, who wondered what would happen to the minority of purposive readers that were traditionally well-served by libraries (Payne, 1983, p. 29):

> In view of this, while the closely classified order may not be the ideal for browsing, but certainly does not make it impossible, the categorized arrangement on the other hand appears virtually to eliminate purposive use of the library. So surveys show that it is only 8% of readers using the library that way; if your readership is but 3000, that is 240 souls. Are they written off? ('If you insist on reading in this infuriating manner, you will just have to move to the next town — or preferably the next county!')

On the other hand, the vision of a singular public in libraries inherited from Cutter's "Rules for a Dictionary Catalog" was not shared by everybody at the time of the reader-interest classifications. Bernard Berelson, for instance, noticed the co-existence of different "publics" in very similar terms as expressed by Olson (Berelson, 1949, p. 63):

> There is no such thing as "the library's public." In the first place, the eligible population of the community is the library's public only in a legal sense, certainly not in a substantive socio-psychological sense. [...] In the second place, the clientele of the library does not represent a public in a meaningful sense. There is no single public of library users; there are several publics. This is not simply a nice semantic distinction. It is central to the problem of the values and the objectives of public library service; it is the key to the 'philosophy of public librarianship.'

In this vein, the different publics envisioned by Berelson included high school students, housewives and white collar workers, business representatives, ambitious young people etc., and a public "of serious-minded people concerned with serious-minded materials on a variety of topics who find in the library what they cannot get elsewhere." It is this public, in the words of Berelson, that should be the main target of public libraries, and therefore of library classifications too.

4.5.2. The Quality and Nature of Categories

Because reader-interest classification meant grouping books together according to the relationships of interest between them, one of the most obvious problems that libraries had to face while defining categories was not only to determine the categories to represent the books, but also how to determine the interests (the relationships) that this singular reader would like to see reflected in the schemes.

Regarding to the categories that are of the interest of the user in terms of representation, this might correspond to the problems of what Hjørland (1992, pp. 175–176) called subjective-idealism "With regard to the user, a document can be ordered with the user's conceptual structures and subject perceptions in mind. The user may well have his subjective grasp of what the subject of the book is [...] he who seeks the key to the concept of 'subject' in the mind of the user commits an error of psychologism." In this vein, among the different approaches that are developed in knowledge organization (Hjørland, 2016), reader-interest classifications would follow the user-based and cognitive approach. The surveys and determination of the "reader's interest" (based on the empirical findings related to those individuals that were surveyed) justified the structure, contents, and the construction of a system that rejected, by principles, all the knowledge and theories of LIS (the "librarians interests"). Currently, the user-based and cognitive approaches in knowledge organization are unsupported. As Hjørland (2013, p. 11) put it, the subjectivity in the construction of knowledge organization systems should be "derived from collective views in discourse communities rather than be derived from studies of individuals or from the study of abstract minds derived from studies of individuals or from the study of abstract minds."

On the other hand, also related to the interests of the users, Reader (1982, p. 34) makes mention of a cartoon by Colin Wheleer that appeared in the "*Library Association Record*" of November '79 — during the peak of popularity of the user orientation movement in the UK, which showed a librarian saying "We're completely user-orientated — we only have books on car maintenance, erotica, and claiming social security benefits." There was clearly some awareness of the perils of going too far in meeting a specific reader's needs, but, Wheeler's cartoon might also be placed within a broader context — one which was framed by the debates over the recreational versus educational role of the public libraries and the materials and categories that should therefore be included in libraries (see Martínez-Ávila & Kipp, 2014, for a further discussion on that matter).

Finally, there was also some suggestion that, in some contexts, categories might have been manipulated in order to dissociate the reader-interest classification projects from the traditional arrangement and philosophy of Dewey. In words of Ainley and Totterdell (1982, p. 125): "In attempting to apply categorisation to the whole stock inevitably one has to make some of the same compromises inherent in a classification scheme and, to a lesser extent, to manipulate readers' interests according to the nature of the stock itself [...] However, it is clear that a better overall arrangement than Dewey is both possible and desirable."

4.5.3. The Problem of Using a Unique Classification System to Reclassify Stock

Reclassifying stock according to the new scheme presented libraries with several problems. Because most reader-interest classifications were developed according to the specific characteristics of each library, there were no mechanisms for interoperability between libraries — or networking — and it was not possible to reuse pre-classified library records. As will be discussed below, this was a factor in the overload of classification work at both local and central levels.

In a traditional case of reclassification — such as DDC-UDC switching — records of other libraries might be consulted that employ the targeted system. But in the case of a reader-interest classification conversion, there were no other libraries using the new scheme and thus no other library records to consult, and all the work of reclassification (and maintenance) of stock had to be done locally from scratch by the library adopting the new system. In addition, the work load in the reader-interest classification projects was considerably increased by other tasks that are inherent to the process of reclassification. As Ainley and Totterdell pointed out (Ainley & Totterdell, 1982, p. 126), the reader-interest classification conversion phase was also accompanied by the clerical tasks of spine-labeling, amending charge cards and date labels, altering the catalog, changing the physical layout, moving stock, re-guiding, informing the public, retraining the staff, etc. At the time that many of these projects were implemented, the use of the Internet, OCLC, Worldcat, cataloging consortia, or other online services to assist with new acquisitions were not used as today, and books usually came to the branch pre-classified according to Dewey by the central library or some other library system service. Yet due to the unique and local characteristics of the new schemes, libraries adopting a local reader-interest classification could not take advantage of any central or collective classification feature for the new acquisitions. Libraries or consortia were not able to use any type of reader-interest classification other than their own, and even in this case records were not made public.

According to Morson and Perry (1982, p. 103), it was the lack of centralization and networking features that led to the failure of reader-interest classification at Ealing Road Library, in the United Kingdom: Because the classification branch was autonomous from the central service, books needed to be reclassified after they arrived at the branch. A second solution for problems commonly associated with the new system was the adoption of hybrid DDC-reader-interest classification models. In these models, both Dewey number and the reader-interest classification categories were kept on the books so libraries could take advantage of the previous classification of the stock and the central pre-classification service for the new acquisitions. But even in these cases, the Dewey part of these dual systems was seen as a theoretical and philosophical burden that had to be kept to a minimum.

When classification and reclassification processes were assisted by the mappings and equivalences of Dewey to the reader-interest classifications, such systems enabled books which were pre-classified according to Dewey to be automatically translated into the new scheme by a quick query of the equivalence, thus facilitating the whole classification process and the clerical tasks that came with it. However, these mappings and equivalences also presented further difficulties upon application such as the need for continuous updating and maintenance of equivalence between both schemes. This was a particularly hard task given the low level of computerization in those days and the lack of standardization in reader-interest classifications. In practice, mappings and equivalences were rarely used beyond the reader-interest classification development process.

4.5.4. The Tension between Standardization and Local Adaptations

One of the most challenging practical aspects that reader-interest classifications had to deal with was the balance between standardization and the needs of local communities. While such schemes were considered to reflect reader's interests in each library, there were no additional mechanisms that enabled such a balance and that also protected the minorities of the libraries. Indeed, the more that libraries embraced a universal standardization (that permitted general interoperability between each other), the more such libraries departed from the specific characteristics of the local communities that supposedly guided the systems. This aspect was noticed by authors such as Denis Chandler of Cambridgeshire libraries, in the United Kingdom: "As more libraries are developed on categorised lines, it is hoped that a balance between uniformity and local need can be maintained so that interchange of books and materials can be streamlined without destroying the response to the special interests of each community" (Chandler, 1982, p. 55).

In fact, lack of standardization was widely regarded as being one of the main causes for the failure in the Detroit Public Library's reader-interest classification project too. As Ainley and Totterdell (1982, p. 129) pointed out:

> The main reason, in fact, for Detroit's failure [...] was that the Reader Interest System was neither centralised nor universal. Only about half the libraries applied the principles with each branch having its own individual categories and notation. This led to such an unnecessary increase in procedures both centrally and at individual branch level, with more staff needed to implement the system, that eventually a drastic cutback in funds prompted the scrapping of this otherwise successful scheme.

Ainley and Totterdell also examined some of the efforts and recommendations of libraries involved in reader-interest classifications projects in order to achieve universality:

> All [libraries reporting reader-interest classification cases] (except Herts and Sunderland) recommend striving for a universal scheme which can be applied to all libraries in the system with its consequent standardisation and centralisation of routine processing. Some go

> even further and suggest that since books can be categorised and/or classified by the selectors at the time of ordering, it is possible to have stock arrive from the supplier fully processed for the shelves. (p. 129)

This latter suggestion, the displacement of the classification process from the library to the supplier had been also proposed in the case of BISAC, where books are pre-classified by the publishers. However, as will be seen later, the proposal in this case was made by the organization in charge of the scheme — the BISG — and not by the libraries involved in carrying out the change.

Among those who were opposed to Dewey, there were some who considered that the system's universality — its "infinite expandability and continuous updating policy" (Ainley & Totterdell, 1982, p. 121) — was also ideal for reader-interest classification. This was not always the case. Indeed, for Ainley and Totterdell, multiple solutions brought additional problems for many libraries seeking universality. This was the case, for instance, with the reader-interest project at Glasgow libraries where there was a "universal reader-interest classification" scheme for all the library system which was made up of a central set of categories, and which had different reader-interest classifications for each branch (which were themselves chosen from the central set of categories). While this system resulted in a certain improvement in local adaptations and in overall interoperability, it also meant that the schemes had to be revised for consistency whenever a category was changed anywhere. In Brent libraries, on the other hand, the branch autonomy of the Ealing Road Library from the central service was considered a disadvantage for all parties concerned: it meant a waste in effort for the central cataloging team and extra work in reclassification at the local level.

In conclusion, the problem of balance between universalization versus local adaptations in reader-interest classification projects had its origins in the fact that every library had a different perception of the needs of their community. As Sapie pointed out, "each library has devised an individualized plan to meet the perceived needs of the community served" (Sapiie, 1995, p. 144) — and in general, all these plans and reader-interest classification measures ended up as individual practices that were hard to standardize and reuse in subsequent projects.

4.6. Reader-Interest Classifications Today: A Global Industry Interest?

While the past cases of reader-interest classifications reported in the literature have been discontinued, the current situation of reader-interest classifications regarding centralization and universalization is different from before, as the global book industry has shown a growing interest in them. For instance, the current E4Libraries Category Headings, developed and maintained by the Book Industry Communications in the United Kingdom, is claimed to to be an attempt to standardize those isolated reader-interest categorizations in fiction and nonfiction over the past 40 years in the British libraries (BIC, 2010). As pointed out in the latest introduction to the "E4libraries Category Headings: Adult non-fiction and adult fiction": "Many library authorities have adopted categorisation schemes for fiction and non-fiction over the past forty years. [...] most of these schemes are unique to the authority concerned and rely heavily for their operation on costly and time-consuming staff intervention."

On the other hand, OCLC is mapping BISAC, the bookstore classification developed by the American BISG, to the DDC. According to the latest "Dewey by the numbers (2016-12-01)" report (http://ddc.typepad.com/025431/2016/12/dewey-by-the-numbers.html), DDC 23 includes 3391 BISAC headings. In addition, libraries adopting BISAC in the United States are having their stock pre-classified via ONYX, replacing the traditional OCLC centralization for the BISG (see Martínez-Ávila, Kipp, & Olson, 2012; Martínez-Ávila, Olson, & Kipp, 2012). However, a study on the use of BISAC in the catalog (Martínez-Ávila, Kipp, & Olson, 2013) revealed that BISAC was not generally being used by all branches of a library system, and when used, BISAC was not being strictly adopted (as provided in the ONIX/MARC records) or being modified in the same way by all branches. This means that BISAC categories (the modifications) are commonly shown in the Call number field of the record and not in the Subjects fields (the BISAC information in the imported ONIX/MARC record was not valid). In consequence, BISAC categories are not retrieved in the Subject searches of library catalogs as they are not included in the proper 6xx MARC fields (Subjects field), and when they are included, these usually are the "canonical" BISAC categories and not the modifications that are used. As for the gathering function of BISAC, this was not being fulfilled in the online catalog either, since no record assigned to a given BISAC category (Call number field) was electronically linked to other records with the same BISAC category unless they shared the same "subjects" in the "Subject field" — usually LCSH. The technical advantages of using a standard for Information Retrieval are being lost in the catalogs because of the necessary modifications of BISAC in libraries; and BISAC is necessarily being modified in libraries because this system is designed for bookstores, entities with very different characteristics. BISAC literals (or their modifications) do not add any familiarity or friendliness to the information retrieval in library catalogs simply because they cannot be retrieved when they are just used as a substitute for Dewey. The advantages of using

BISAC for browsing and the physical organization of books in libraries are also arguable, but if BISAC categories cannot be retrieved, the use of BISAC in online catalogs is questionable.

On the other hand, centralization and universalization in library standards are not exempt of problems and criticism. In the words of Fox and Olson (2012, pp. 86–87):

> centralization in cataloging as practiced by OCLC, however, has turned out to be more ominous in the power exerted over constituents/customers. [...] While Egan and Shera may advocate centralization with the assumption that all stakeholders have a say in influencing the resulting knowledge structures, the danger is that standardization provides the underlying justification for a goal of universality, thus thwarting the standpoint perspectives [...] even the oxymoronically named "WorldCat Local" avoids allowing any decision-making or personalization at a local level

While the local focus of reader-interest classifications did not include the voice of marginalized groups, with the exception of only one paper (Bridgwater, 1990), current (local) reader-interest classifications practices do not seem to deviate too much from the global OCLC activities and problems that the old Dewey represents. For instance, the development of BISAC is dictated by the pressure of sales logic and lobbyists such as the Evangelical Christian Publishers Association (ECPA), and driven by a "market warrant" (Martínez-Ávila & Kipp, 2014). In this vein, the main difference between the old cases of reader-interest classifications and BISAC would be the approach that is followed. While reader-interest classifications followed a cognitive and user-based approach, it could be argued that BISAC follows a practicalist and intuitive approach (the same approach that the DDC follows, by the way, in spite of being presented as an alternative), as it was devised as a practical way to exchange and organize information according to the industry convenience (Martínez-Ávila, 2016).

As for the problems that affected reader-interest classifications in the past, while universalization has been achieved in current reader-interest classifications, the promised advantages of centralization have not. In this not so positive scenario some questions remain: Can libraries' independence rely on OCLC's DDC-BISAC mappings? What happens for instance to reclassification with each new version of BISAC? Since new books will be always pre-classified according to the latest version of the scheme, libraries cannot rely on the centralization practices of a book industry in which bookstore stock is sold/gone while the library collection is not. In this sense, while there is not an easy solution, libraries that consider adopting a reader-interesting classification must really think of the interest of the users (in plural) and not only of the global book industry that dominates the development of the standards.

References

Ainley, P., & Totterdell, B. (1982). *Alternative arrangement: New approaches to public library stock*. London: Association of Assistant Librarians.

Berelson, B. (1949). The implications for library policy. In *The library's public: A report of the public library inquiry*. New York, NY: Columbia University Press.

Betts, D. (1982). Reader interest categories in Surrey. In P. Ainley & B. Totterdell (Eds.), *Alternative arrangement: New approaches to public libraries stock* (pp. 60–77). London: Association of Assistant Librarians.

BIC. (2010). *E4Libraries category headings – Adult non-fiction/adult fiction*. Retrieved from http://www.bic.org.uk/files/pdfs/101004%20E4Libraries%20Category%20Headings%20-%20ANF%20and%20AF%20-%20revised.pdf

Bridgwater, S. (1990). Out of the doldrums and into the curriculum: De beauvoir junior school library. *The School Librarian*, *38*, 53–54.

Chandler, D. (1982). Self-service libraries: Providing for the smaller community in Cambridgeshire. In P. Ainley & B. Totterdell (Eds.), *Alternative arrangement: New approaches to public libraries stock* (pp. 47–59). London: Association of Assistant Librarians.

Donbroski, L. (1982). Categorisation at east Sussex county library. In P. Ainley & B. Totterdell (Eds.), *Alternative arrangement: New approaches to public libraries stock* (pp. 78–85). London: Association of Assistant Librarians.

Donovan, J. M. (1991). Patron expectations about collocation: Measuring the difference between the psychologically real and the really real. *Cataloging & Classification Quarterly*, *13*(2), 23–43.

Fox, M., & Olson, H. A. (2012). Feminist epistemologies and knowledge organization. In R. P. Smiraglia & H.-L. Lee (Eds.), *Cultural frames of knowledge* (pp. 79–97). Würzburg: Ergon.

Hjørland, B. (1992). The concept of 'subject' in information science. *Journal of Documentation*, *48*(2), 172–200.

Hjørland, B. (2013). User-based and cognitive approaches to knowledge organization: A theoretical analysis of the research literature. *Knowledge Organization*, *40*(1), 11–27.

Hjørland, B. (2016). Knowledge organization (KO). *Knowledge Organization*, *43*(6), 475–484.

Hubbard, L. (1972). Libraries for an age of exploration: A reader interest approach. *The Unabashed Librarian*, Spring, 26–28.

Martínez-Ávila, D. (2016). BISAC: Book industry standards and communications. *Knowledge Organization*, *43*(8), 655–662.

Martínez-Ávila, D., & Kipp, M. (2014). Implications of the adoption of BISAC for classifying library collections. *Knowledge Organization*, *41*(5), 377–392.

Martínez-Ávila, D., Kipp, M., & Olson, H. A. (2012). DDC or BISAC: The changing balance between corporations and public institutions. *Knowledge Organization*, *39*(5), 309–319.

Martínez-Ávila, D., Kipp, M., & Olson, H. A. (2013). Uso de BISAC como sistema de organización del conocimiento en el catálogo. In V. Dodebei & J. A. Chaves Guimarães (Eds.), *Complexidade e organização do conhecimento: desafios de nosso século* (pp. 195–199). Rio de Janeiro: ISKO-Brasil.

Martínez-Ávila, D., Olson, H. A., & Kipp, M. (2012). New roles and global agents in information organization in Spanish libraries. *Knowledge Organization*, *39*(2), 125–136.

Martínez-Ávila, D., & Satija, M. P. (2015). Reader-interest classification as a method of stock control: The McClellan legacy. In J. A. Chaves Guimarães & V. Dodebei (Eds.), *Organização do conhecimento e diversidade cultural* (pp. 135–142). Marília: ISKO-Brasil.

Martínez-Ávila, D., & Segundo, R. S. (2013). Reader-interest classification: Concept and terminology historical overview. *Knowledge Organization*, *40*(2), 102–114.

Martínez-Ávila, D., Segundo, R. S., & Olson, H. A. (2014). The use of BISAC in libraries as new cases of reader-interest classifications. *Cataloging & Classification Quarterly*, *52*(2), 137–155.

McCarthy, A. (1982). Burning issues: Stock appeal in Sunderland. In P. Ainley & B. Totterdell (Eds.), *Alternative arrangement: New approaches to public libraries stock* (pp. 88–100). London: Association of Assistant Librarians.

Morson, I., & Perry, M. (1982). Two-tier and total: Stock arrangement in Brent. In P. Ainley & B. Totterdell (Eds.), *Alternative arrangement: New approaches to public libraries stock* (pp. 101–118). London: Association of Assistant Librarians.

Olson, H. A. (1997). Thinking professionals: Teaching critical cataloging. *Technical Services Quarterly*, *15*(1–2), 51–66.

Olson, H. A., & Schlegl, R. (2001). Standardization, objectivity, and user focus: A meta-analysis of subject access critiques. *Cataloging & Classification Quarterly*, *32*(2), 61–80.

Payne, R. (1983). Fears on categorization. *Library Association Record*, *85*(1), 29.

Reader, D. (1982). User orientation in a Hertfordshire branch. In P. Ainley & B. Totterdell (Eds.), *Alternative arrangement: New approaches to public libraries stock* (pp. 34–46). London: Association of Assistant Librarians.

Rutzen, R. (1952). Shelving for readers. *Library Journal*, *15*, 478–482.

Sapiie, J. (1995). Reader-interest classification: The user-friendly schemes. *Cataloging & Classification Quarterly*, *19*(3–4), 143–155.

Satija, M. P. (2004). *A dictionary of knowledge organization*. Amritsar: Guru Nanak Dev University.

Sawbridge, L., & Favret, L. (1982). The mechanics and the magic of declassification. *Library Association Record*, *84*(11), 385–386.

Chapter 5

Knowledge Representation of Photographic Documents: A Case Study at the Federal University of Pernambuco (Brazil)

Fabio Assis Pinho

Abstract

This research demonstrates that the combination of two methodologies to describe photographs — Nobrade and Sepiades — together with the contextualization and identification of the content informational model of the photograph facilitates the reconstruction of institutional memory. This exploratory research based on case study adopted a set of metadata from Nobrade and Sepiades aiming at the organization and availability of the set of information extrinsic to the 20 portraits of the rectors of the Federal University of Pernambuco from 1946 to 1971. For the contextualization of this period, some of the events that occurred in Brazil were highlighted, as they influenced the academic environments during the Democratic Period and the Military Regime. The description of the photography was made in four parts: first part was called administrative dates resulting in the photo identification information; second part was the provenance data, that is, data about its origin and context; third part was composed of the technical data of the photograph; and fourth part was composed of the image data which contains information about the content of the photograph. The second and fourth parts using context and content information enable the photograph to be understood beyond what can be seen and contribute to the reconstruction of institutional memory. This research contributes to the elaboration of a

The Organization of Knowledge: Caught Between Global Structures and Local Meaning
Studies in Information, Volume 12, 71–86

ISSN: 2055-5377/doi:10.1108/S2055-537720170000011011

documentary system using a combination of methodologies, focusing on photographs, not only as an institutional technical activity, but also as an activity necessary for the reconstruction of the institutional memory.

Keywords: Information science; knowledge organization; institutional memory; photography; indexing; knowledge organization system

Chapter Outline

5.1. Introduction

The set of procedures involved in the document management of memory institutions begins with the activity that includes physical and content organization, the identification of the availability to the public, and the research instruments resulting from this representation. These activities facilitate information retrieval.

The documentation process done at memory institutions includes a set of activities that organizes documents, from their identification to their location, conservative treatment, and disposal. In the context of documentation, according to Le Coadic (2004, p. 15), it "is a set of techniques to organize, analyze documents, describe and summarize them." This process of organization and analysis is limited not only to books, but also to any documental type.

In this respect, photographic documents that require specific methods not only for packaging and preservation but also for your informational organization and representation are also included.

Photographic documents require specific methods for representation, especially regarding their information content. Regarding their cataloging, the most widely used models are ISAD (G) — International Standard Archival Description (General), ISBD — International Standard Bibliographic Description, AACR2 — Anglo-American Cataloguing Rules 2nd edition, Sepiades (SEPIA Data Elements Set), and Dublin Core. As a consequence of this variety of ways for the description, there might be problems for information retrieval and the construction of memory.

The photographic collections are sometimes represented in various forms, using, for example, the AACR2, the Brazilian Standard for Archival Description — Nobrade, and also the European Programme for Stock Photography Images Safeguarding Access — Sepiades. So the results of the descriptions, especially regarding their contents or contextual information, may vary.

The problem of how varied the photographic documentation is described will affect the institutions that deal with the particular implementation of methodologies designed for textual collections, which do not address the specificities of the photographic document, and, particularly, the technical information. These methods result in a factor that deserves mention, namely, the set of elements to describe the photos.

These elements are often difficult to identify, for example, while describing a picture and understanding or even interpreting what is seen in the photographic document. Furthermore, this description is restricted only to what the image shows, what is called by Manini (2002) as informational content, or semantic content, which is information shown in the image, at the expense of nonvisual information.

The informational content of the photograph allows its contextualization so that people who were not part of the image can understand it beyond what is seen in the image. In this sense, we have the following research question: does a field of description of the informational content of photography allowing its contextualization contribute to the construction of institutional memory?

Empirically, these issues can be noted in the photographic collection, under the custody of the Denis Bernardes' Memorial, which is a space created by the Department of Information and Communication Management (Segic) of the Federal University of Pernambuco and its Central Library. This space has been installed in order to preserve the institutional memory of the university through the storage and scanning of documents.

The name of the Memorial is a tribute to Professor Denis Bernardes, who served in the Department of Economics and Social Services and highlighted the interest in the preservation of University documents. He was a historian and his latest research dealt with the relationship between memory, information, and society. Currently, the collection consists of about 15,000 photographs related to the activities of this university, developed over the years, including its history in images with various topics.

This photographic collection consists of commemorative events, records, protests, election of rectors, etc., which are the documentation of its memory. Analyzing this photographic documentation is a challenge and contribution to the record of memorials aspects of this institution. Henceforth, this photographic collection serves as a corpus for research and activities involving undergraduate students and master students at Federal University of Pernambuco.

In this vein, the hypothesis here is that the absence of a field allowing the contextualization of the photographic image negatively affects the retrieval of the information and the construction of memory through the collection that the photography belongs to. Thus, it is understood that the possibility of portraying the world also associates the photography to their constitution as a memory element, especially when represented in local systems.

In this sense, it is sought to enable a field of description of the photographic document content that allows its understanding and contextualization. The justification here is that the photography leads the transmission of memory to the background; the occasion or the people who were photographed reveal some context that it is not understandable for those who were not involved in the photographed event.

Therefore, the contribution of this research lies in the fact that it allows an approximation of the knowledge representation of photographic collections to the institutional memory through descriptive fields of informational content.

As a case study a collection of photographs belonging to the Federal University of Pernambuco, the Denis Bernardes' Memorial, was also subject to documentary analysis in such a way that its content was contextualized in a local system, in spite of being included in global systems. In this sense, it was found that for the application of documentary analysis to the photographic collection it was also necessary to consult other documents for the contextualization of the photography and thus to contribute to the construction of institutional memory.

Therefore, the main objective of this research was to demonstrate the need of the contextualization of photographic documents to include them in local systems, in order to allow the construction of institutional memory.

5.2. Institutional Memory and Knowledge Representation of Photos

Institutions are understood as the locus of knowledge and power in a Foucauldian sense and have two faces: to remember and to forget. Memory as a political issue passes through certain discourses and is allied to the criteria of truth. In this sense, the institutions are what we collectively manage and, therefore, connect to the memory. For this reason, the concept of institutional memory is hybrid. Also for this reason, the concept of institutional memory has one main element: information.

The concept of institutional memory is related to memories and the validation and storage of what is socially constructed. The set of documents produced in an institution reflecting its activities and history seeks the systematization and availability of information to society, so that the institution preserves the memorial records of those who contributed to its development and reflects the individual and collective memory. Among the theorists who work with the concept of memory, we can highlight Maurice Halbwachs, Jacques Le Goff, and Michael Pollak among others.

The issue of institutional memory shows special characteristics in the process of institutionalizing social practices, for example, inaccessible files are open to users in a "power game" because truth is time-dependent. The institutional memory is a permanent set of information that is built upon dynamic discursive practices that, in turn, are materialized in documents (including photography). The concept of institutional memory is composed of information, memory, and the institution that acts in the process of institutionalization of social relations.

In this sense, institutional memory is the reconstruction of the past and the institution's use of its own history as a set of elements that confirm its identity. This can be done using documents, recorded facts, and photographs (Bergson, 2006; Worcman, 2004).

The use of the photography as a research source and historical document has been established as an important instrument in the process of the reconstruction of institutional memory, because it is a cutout of a scene and a given aspect of the world. Through characteristics such as luminosity, intensity of the variation of tones, spatiality, and temporality, the image forms and produces knowledge. It denotes sensitive qualities that are inherent to the natural world and that can be perceptually and cognitively interpreted by people.

In this aspect, the photographic document retains data that is inherent to the moment that they reflect, that is, they bring hints or evidences to retrieve facts or events that originated them.

And what is the importance of photography for the reconstruction of institutional memory? According to Kossoy (2007, p. 41), it is the indication that exists in the image (iconographic) which in addition to historical, geographical, geological, and anthropological elements carries meaning. A set of written and visual information that, related to each other, allows us to date, geographically locate, identify, and finally retrieve micro-histories of different natures that are implicit in the document.

The purpose of recording and storing documents, as a means not to forget and get memory lost, is also present in the photography. Regarding this, Sontag (1981, p. 4) points out that:

> To photography is appropriates of the thing photographed. It engages in a certain relationship with the world that resembles the knowledge — and therefore with power. [...] The picture plays with the scale of the world, can be reduced, enlarged, cut, cut, fixed and distorted. Ages to be infested by the common diseases to made of paper objects; fades away; we value, it is bought and sold; It is played.

This can be evidenced in the way people take ownership of the camera. There is a great willingness to record the times that were considered important and to show them to others, as emphasized by Santos Júnior (2008, p. 7): "In fact, the power of the camera led people to believe that the time is to interesting events worthy to be photographed. And if they are worthy of being photographed, it is not strange that all are happy to show your photos to others, especially the records of tender and loved ones."

This possibility of portraying the world in its various sizes associates to photograph their constitution as the memory element. This, perhaps, is the way to do "forever." This is the great value belonging to the picture. The photograph "revolutionized memory" because, immediately, the picture can activate memory, talk about a past, allowing revive it in the present, not only being belonging to the individual who observes, but also to recall from its past (Le Goff, 2003).

Perhaps the clearest example to understand this claim is the production of family albums. The memory consists of creation and recreation of images all the time. There is no gap between present and past; there is a mix of memory; memory and fiction revived at the time of recall. The memory refers to a past event born of this emotion.

According to Monego and Guarnieri (2012, p. 73):

> The photo is taken to support memory. Whereas memory is a living process that each of us gets through time. These lived [...] experiences are a kind of equity that are used alone and passed on to other generations through the memories. An important factor in memory is the picture, as it works as a kind of social memory, capable of recording moments, people and places that never exist. In this sense, the photograph has been used as a form of reconstruction of memory, both as an individual and as a member of various social groups.

Then, the picture is subject to recorded memory and also to its use and construction as product of socially constructed memory. The photograph constitutes an object of memory; it is a cutout of a time and space, which brings with it a past event. The picture is capable of recording memory and also using and constructing a product of socially constructed memory.

The photograph at the time it is taken is already a past object that is stored in personal or institutional archives. Thus, it constitutes one of the people memory elements. Through it, people can redeem past moments, no longer existing places, untold stories, individual and collective lived memories, etc. This is one of the roles assigned to photography in its most 150 years of existence (Manini, 2002).

Over the years, during its development and improvement, the photography always had a veracity character, used as a source attesting the facts. In this sense, Kossoy (1999, p. 19) mentions that it has been accepted and used as a definitive proof, witness to the truth, though its real representation feature does not give the picture the absolute credibility.

In this regard, Barthes (1984, p. 20) points out that the picture is constituted as a representation of the real trial, and, before that, it can be the object of three practices, namely do, support, and look (or three emotions or intentions).

The photographer is the person who sets the scene for the *spectator*, equivalent to his/her look. The support is the referent, the object, the subject, the show which was held in the photographic emulsion. Thus all photographic frames are composed of an *operator* (photographer), *spectrum* (look), and *spectator* (observer), and to understand them is necessary to unravel the operator context represented by a *spectrum* that will turn into a *spectator*.

A peculiar feature of the photography is the irreversibility of the recorded time. Thus, the photography guides the placement of memory either individually or collectively. That's what makes, for example, that the picture is present in our life events. The moments that were selected deserve to be remembered and become motives of the shots of our cameras. Therefore, the setting, the time, and the people who were photographed reveal some context, in a way that their representation is just not quite understandable for those who were not involved in the depicted event. Therefore, it is necessarily a more thorough understanding and contextualization.

In an attempt to deal with the context of the image, it is important to understand and reveal the unseen parts of the photographed image, which are elaborated and specific methods for its treatment and retrieval of its contents. Knowledge representation is proposed to study the methods for this, seeking a theoretical and methodological framework suitable for the intrinsic aspects of the photographic document that can also be retrieved.

Therefore, it is considered that the contextualization of the photographic image needs support in exogenous documentation so that the image can be represented adequately in relation to its content and be included in local and global systems.

The photographs, besides illustrating texts, construct a narrative that is independent, drawn from the analysis and interpretation of the information contained in the document itself. For the understanding of the photographic document, it is necessary to analyze this in addition to the texts (Leite, 1993).

The challenge for the representation of photographic documents is to understand the pictures on their production time, and inquiries should go beyond its support. As stated by Leite (1993, p. 72): "questions should go beyond poses, framing, clothing and costumes, or type, size, date and location of the works. It is essential that the photo analysis also considers what was hidden, as opposed to what it is emphasized, the feelings and representations that do not transpire easily through these documents, and also the forms of acceptance, use and circulation of this type of image."

To analyze a photograph is not only using the imagery data, but also using information obtained from various sources: the title, the legend, notes on the back of the photo or from other documents — manuscripts in general — accompanying the picture, representing a great support to the identification (Manini, 2002).

One of the problems that is commonly present in the documentation of a photographic collection is that it has no associated identification, making it difficult to contextualize it. Identifications, when they appear, are just inscriptions on the back or on labels stuck on the medium, and even when they present, the information is not always related to the imagistic content; sometimes this information is circumstantial (Manini, 2002).

This circumstantial information is no less important as it allows the photo identification of the boot, because with it we can try to make the journey in reverse. The presence of stamps, dedications, institutional names is very common, and then the need to make a withdrawal from this referenced information.

Another factor that stands out in the treatment of photographic documents is that in some cases it includes methodologies and practices that are applied to written records, which brings in itself some problems because photographic documents cannot be interpreted and analyzed as text documents as they show the specificities of another document type.

Thus, the complexity of the genesis, analysis, and understanding of photographs should be systematized, resulting in focused methodologies and practices that are exclusive for that document type. The information that makes up the document is within the holder, and also external to them; therefore, all photographic records are a piece of reality, formed by intrinsic and extrinsic information (Kossoy, 1999).

So we need an appropriate methodology for the thematic representation of the photography with emphasis on the collection of information related to the content. This is an information related to the context in which they were produced, emphasizing the recording of events during their production. Tables 5.1–5.3 are examples of methods to representation of photographic documents for information systems.

Table 5.1: Systematization of the methodology proposed by Panofsky in 1979.

Pre-Iconographic	Iconographic	Iconological
Description of the elements shown	*Identification of the elements shown*	*Understanding and conceptual analysis of the image, considering the elements identified in previous levels*

Source: Panofsky (1979).

Table 5.2: Systematization of the methodology proposed by Manini in 2002.

	Informational Content			Expressive Dimension
	From		About	
Category	Generic	Specific		
Who/What				
Where				
When				
How				

Source: Manini (2002).

Table 5.3: Systematization of the methodology proposed by Rodrigues in 2007.

Physical Description	Composition	Archival Context	Denotation	Connotation	Aboutness

Source: Rodrigues (2007).

These three tables are examples of photographic image analysis that extract their informational content beyond what is seen.

When analyzing images, the discovery of information passes by wandering its surface. In this sense, Flusser (2002, p. 8) argues that anyone who wants to "deepen" the meaning and restore the abstract dimensions must allow his or her view to wander around the surface of the image. Such wandering by the surface is called scanning. This closer look at the photography convexly reveals its symbols, while the look of the researcher establishes temporal relations between the elements.

Based on the analysis of the systematizations from the tables, it is found that the general procedure of document content analysis requires the establishment of a conceptual framework to guide the process and to include the understanding of its context of communication, analytical procedures, and documentary standardization of instruments (Agustín LaCruz, 2010). Considering these aspects, Agustín LaCruz (2010) presents a procedure for pictorial image content analysis, distributed in four operations, which can be seen in Figure 5.1.

Determining the content of the images in operation consists of three distinct phases: description, identification, and interpretation. These phases include the three planes of the meaning of work: the pre-iconographic one, the iconographic one, and the iconological one, considering the theme, characteristics, gender, usage, style, context, etc.

The description is the most basic step. It includes the identification of the topics presented in the images, featuring the objects and being represented. The identification process comprises an individual and detailed analysis of which was identified in the previous phase.

According to Agustín LaCruz (2010, p. 101), this phase is supported by another documentary task: the phase of the exogenous documentation and

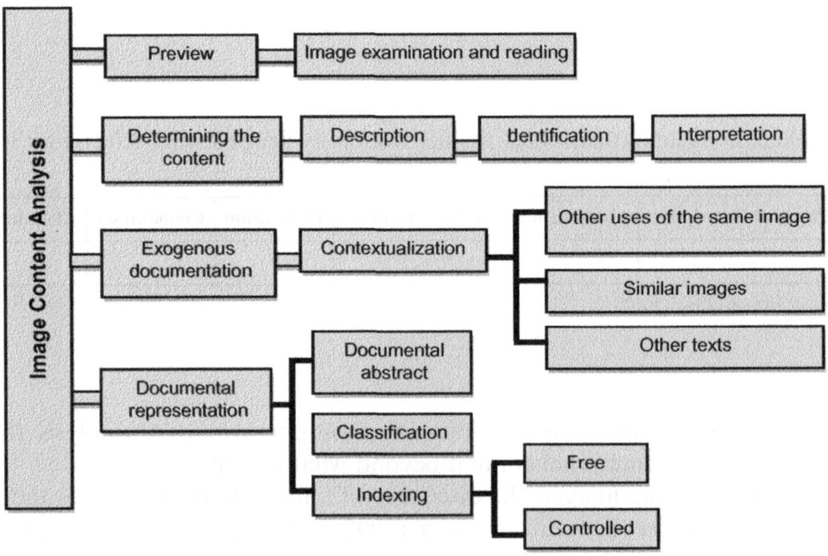

Figure 5.1: Procedure for image content analysis. *Source*: Agustín LaCruz (2010).

location of sources of information, which consists of the consultation by the researcher or information officer of other works, biographies, and dictionaries.

The interpretation includes the deepest level of the process, the explanation of the image content, which is based on the context.

The exogenous documentation consists of the analysis of a number of sources outside the imagistic document analyzed and can contemplate primary, secondary, textual, and imagistic sources. Based on this exogenous documentation we can contextualize the photographs that are being organized and described.

And finally, the fourth stage includes the document representation that generates the documentary products: indexing, classification, and summaries which enable the representation of documents. This step is the organization of information to be made available in databases for the users.

These levels explain that the analysis of the photographic image must be deeper than what is shown, as history shows that there is innumerable knowledge that takes years to emerge and flourish.

In particular, knowledge representation (indexing, cataloging, and classification) can help in the reconstruction of the institutional memory using the photographs. This is possible because a photograph can be contextualized using its informational content, a content that goes beyond what is seen in the photography.

5.3. Knowledge Representation of Photos at Federal University of Pernambuco

The specific methodologies for this document type are designed to organize the information in the pictorial media. The photos are documents that professionals who work in memory institutions — museums, archives, and libraries — started to develop focusing on this practice, highlighting the need to systematize the set of information that supports to promote it and make it available to the public and their members.

These facts have caused researchers to reflect a lot on the existing photographic representations, looking for methods and means to decipher them in order to obtain a rich contribution to meanings and information.

Following this trend, the Federal University of Pernambuco has also begun to organize his photographic collection, so their treatment and representation can assist in setting up an institutional memory. Thus, the diagnosis, classification plan, packaging, representation, and preservation of photographic collections were developed. Below, there is an example of how the representation of a photographic document and how this representation can assist in the institutional memory of the university.

At that time, most attention was put on the use of methodologies for another document type, so the new procedures for knowledge representation of photographs also consider the components of Nobrade and Sepiades. As a consequence, this chapter discusses the problems about its description and representation using documentation and its reflection on the institutional memory.

The methodological approach adopted here is the application of Nobrade and Sepiades, based on the elaboration of a set of metadata aiming at the organization and availability of the set of information extrinsic to the photographic documents. To demonstrate this correlation, the combination of Nobrade and Sepiades was applied to the portraits of the rectors of the Federal University of Pernambuco from 1946 to 1971.

The research corpus consists of 20 portraits of university rectors, from 1946 to 1971, comprising the rectorates of Joaquim Ignácio de Almeida Amazonas, João Alfredo Gonçalves da Costa Lima, and Murilo Humberto de Barros Guimarães. For the contextualization of this period, some of the events that occurred in Brazil were highlighted, as they influenced the academic environments during the Democratic Period and the Military Regime.

The identification and recording of these events were organized and contemplated in the description stage of the photographic corpus, which was carried out from the set of metadata resulting from the interaction between Nobrade and Sepiades, according to Table 5.4. The exogenous documentation used to contextualize the photograph was composed of minutes of meetings and internal newspapers.

Table 5.4: Example of knowledge representation of photography for institutional memory.

Administrative Data	
Code	BR UFPE IU PF.R.R 2389
Type of code	Assigned by Professional
Related code	Not identified
Related unit	Not identified
Date	Not identified
Dimension	18 × 24.3 cm
Place of collection	Knowledge Technologies Laboratory — Líber
Receiving guide	Not identified
Receipt date	2013
Diffusion	Not identified
Reference of diffusion	Not identified
Described by	Ana Cláudia Santos and Conceição Wanderley
Description date	November 18, 2013
Provenance Data	
Background	Federal University of Pernambuco
Group	University Press
Series	Photography
Subseries	Portraits
Dossier	Portraits of the Chancellors 1946–1971
Note of context	João Alfredo da Costa Lima was the second chancellor of the University of Recife (UR). He took the office after the death of the former chancellor Joaquim Amazonas. His administration included the period of 1959–1964 and was marked by internal problems, incited by Professor Gilberto Freyre that resulted in his resignation two months to his conclusion. Events: reform in the Rules and Regulations of UR; student strike at the Law Faculty; process to change the name of the campus to Campus Joaquim Amazonas; and creation of Research Centers and Institutes of Education and Research. After the resignation, the UR was managed by the vice-chancellor Newton Maia, who organized a new electoral process for the next chancellor — Murilo Guimarães
Technical data	
Photographer	Not identified
Type	Photo
Documentary tradition	Original
Support Date	Not identified
Material	Paper
Format	Horizontal
Technic	Negative film
Dimension	18 × 24.3 cm
Sharpness	Great
Color	Black and White
Polarity	Positive
Notes	It is written on the back of the picture: João Alfredo, 2389

Conservation	Good conservation. The photograph has dirt on the edges, with handwritten entries in pencil and pen. It was made preventive maintenance procedure by mechanical cleaning with the use of rubber powder. The packaging and storage were carried out in polyester jacket and acid free box
Image data	
Image title	Event with the chancellor
Type of the title	Assigned by Professional
Date of the image	June 1, 1959 to June 1, 1964
Place of the image	Pernambuco — Recife
Character	João Alfredo Gonçalves da Costa Lima
Expressive dimension	Wide-angle, snapshot, artificial light, daytime, general view, close, focal length, portrait with depth
Related terms	Portrait, João Alfredo
Informational content	Professional portrait, half body, containing three male figures standing. From left to right, John Alfredo wears a suit in dark color, black tie, and white shirt. He wears glasses and his head is slightly tilted to the left side, facing the viewer. In the center is an unidentified man wearing suit in dark color, black tie, and white shirt. He's the head turned to the right and crack a smile on his face. On the right side has another unidentified man wearing glasses and dark suit, white shirt, and black tie, his head directed to the first portrayed, that is before him. On the third level, to the right, inner space contains an iron railing

Table 5.4 shows an example of how the combination of Nobrade and Sepiades resulted in the description of the photographic document. The first part was called administrative dates resulting in the photo identification information. The second part is the provenance data, that is, data on the origin and context of the photograph. The third part is composed of the technical data of the photograph. The fourth and last part is composed of the image data which contains information about the content of the photograph. In the latter part, the methodologies presented in Tables 5.1–5.3 were taken into account.

The second and fourth parts using context and content information enable the photograph to be understood beyond what can be seen. Because of this, the contribution to the reconstruction of institutional memory begins.

This example shows the concern with the increasing number of photographic documents belonging to Federal University of Pernambuco and memory institutions that has to deal with a differentiated support that is commonly treated. Based on this need, several studies have been conducted and are being developed exclusively for this document type.

The note of context and informational content allows the construction of institutional memory based on the knowledge representation of these photographs. The administrative, technical, and image data allow us to understand the photographed time. Furthermore, this knowledge representation allows these documents from local systems to be included in global systems.

Thus, organizing and representing photographs is necessary because of the impossibility of accessing to all this production by the users, as well as its importance to the institutional memory. In addition, memory institutions have photographic collections that begin to consider the specificities of the photographic content, however, this is of great importance when the considered aspects are not part of the recorded of information, something to lead to their loss.

This chapter contributes to the elaboration of a documentary system using a combination of methodologies, focusing on photographs, not only as an institutional technical activity but also as an activity necessary for the reconstruction of the institutional memory. This also contributes to the reflection on the practice of the documentation adopted in the process of description and representation of the photographs considering their extrinsic information.

5.4. Conclusion

In conclusion, we emphasize the importance of the photographs for the creation of institutional memory and how this type of document needs to be properly represented and organized.

A collection of photographs belonging to the Federal University of Pernambuco, through the Denis Bernardes' Memorial, was subject to documentary analysis in such a way that its content was contextualized in a local system, in spite of being included on global systems. In this case, we believe that the knowledge representation of photographs was satisfactory for the reconstruction of institutional memory.

Moreover, we believe that the contextualization of the photos was necessary for the reconstruction of institutional memory as shown in the example shown in Figure 5.1, and supported by the theoretical framework mentioned above.

In addition to this, the approach presented here does not include a historicizing action of the Federal University of Pernambuco, as for this it would be necessary to carry out a systematic survey of other documents, not just the photos. This is just a reflection on the knowledge representation of photographs for institutional memory.

References

Agustín LaCruz, M. C. (2010). El contenido de las imágenes y su análisis en entornos documentales. In R. Gómez Díaz & M. C. Agustín LaCruz (Eds.), *Polisemias visuales: aproximaciones a la alfabetización visual en la sociedad intercultural* (pp. 85–116). Salamanca: Universidad de Salamanca.

Barthes, R. (1984). *A câmara clara*: notas sobre a fotografia. Rio de Janeiro: Nova Fronteira.

Bergson, H. (2006). *Matéria e memória*: ensaio sobre a relação do corpo com o espírito (3rd ed.). São Paulo: Martins Fontes.

Flusser, V. (2002). *Filosofia da caixa preta*: ensaios para uma futura filosofia da fotografia. Rio de Janeiro: Relume Dumará.

Kossoy, B. (1999). *Realidades e ficções na trama fotográfica*. São Paulo: Ateliê Editorial.

Kossoy, B. (2007). *Os tempos da fotografia*: o efêmero e o perpétuo. São Paulo: Ateliê Editorial.

Le Coadic, Y-F. (2004). *A ciência da informação*. Brasília: Briquet de Lemos.

Le Goff, J. (2003). *História e memória*. (5th ed.). Campinas: Ed. UNICAMP.

Leite, M. M. (1993). *Retratos de família*. São Paulo: Edusp.

Manini, M. P. (2002). *Análise documentária de fotografias*: um referencial de leitura de imagens fotográficas para fins documentários. Dissertation (PhD in Communication Science), Universidade de São Paulo, 2002.

Monego, S., & Guarnieri, V. (2012). A fotografia como recurso de memória. *Revista Cadernos do CEOM*, *25*(36), 71–87. Retrieved from http://bell.unochapeco.edu.br/revistas/index.php/rcc/article/view/1153

Panofsky, E. (1979). *Significado nas artes visuais*. (2nd ed.). São Paulo: Perspectiva.

Rodrigues, R. C. (2007). Análise e tematização da imagem fotográfica. *Ciência da Informação*, *36*(3), 67–76.

Santos Júnior, N. B. (2008). *Fotografia e memória*: contra a ação do tempo, foto fortalece tradição das técnicas de memorização. In IV Congresso Internacional de Comunicação, Cultura E Mídia. Retrieved from http://www.belasartes.br/revistabelasartes/downloads/artigos/1/revista-ba-foto-memoria.pdf

Sontag, S. (1981). *Ensaios sobre a fotografia*. Rio de Janeiro: Arbor.

Worcman, K. (2004). Memória do futuro: um desafio. In P. Nassar (Ed.), *Memória de empresa: história e comunicação de mãos dadas, a construir o futuro das organizações* (pp. 20–25). São Paulo: Aberje.

Chapter 6

Slanted Knowledge Organization as a New Ethical Perspective

José Augusto Chaves Guimarães

Abstract

The purpose of this chapter is to characterize knowledge organization (KO) as a field that is affected by geographic and diachronic variations in such a way that the recognition of a *slanted KO* could be considered an ethical option in the KO theory and practice. KO can be considered a dynamic social product that reflects a construction that is altered in space and time. Slants are inherent to any organization of knowledge and are manifested in multiple dimensions. There is a need to find a balance between the respect for the local specificities and the necessity of global access to information. Conceptual and terminological time and space slants in KO are presented. Examples of possible day-by-day searches are analyzed in order to evidence the different cultures that are involved in the different social-linguistic characteristics. The recognition of time and space as operational axes for an ethical approach to a slanted KO is important because: (a) it tries to intervene in represented and possibly disseminated biases that are practiced so far; (b) it recognizes the coexistence of diverse groups and communities, with local characteristics, meanings, and idiosyncrasies, that will need to communicate with each other in global information systems of information; and (c) it can promote an intercultural ethics of mediation, culturally warranted, in order to avoid cultural damages and to guarantee that descriptions can reflect

The Organization of Knowledge: Caught Between Global Structures and Local Meaning
Studies in Information, Volume 12, 87–102

ISSN: 2055-5377/doi:10.1108/S2055-537720170000011012

the past while keeping an eye in the future, based on KOS whose functionality remains over time.

Keywords: Ethics in KO; prejudices; proselytism; space and time in KO; bias; slant

Chapter Outline

6.1. Introduction

Although the Information Science literature points out the pioneer work of Bliss in 1931 as the "birth" of the knowledge organization (KO) field, it is only four decades later, with the studies of Soergel and Dahlberg in the 1970s, that KO assumed a scientific dimension, an aspect that was materialized in the creation of the International Society for Knowledge Organization (ISKO), in 1989.

Standing on an interdisciplinary ground with disciplines such as Neuroscience, Psychology, Epistemology, Information and Communication Sciences, Semiotics, Linguistics, Mathematics, Logic and Computer Science (García Marco, 1997), KO has been reflecting on the challenges of a world that shows an ever-increasing production of scientific knowledge and a less effective access to it (Jaenecke, 1994).

In this sense, many efforts have been carried out in order to discuss the scientific status of the KO field, especially in the discussion of its conceptual dimension (Andersen, 2002; Café & Brascher, 2011; Dahlberg, 1993, 2006; Furner, 2009; Gilchrist, 2011; Gnoli, 2008, 2011; Hjørland, 2008, 2009; Jaenecke, 1994; Kiel, 1994; López-Huertas, 2008; Tennis, 2008; Zins, 2004; among others), whose object — knowledge — stands in a *razor's edge* between an individual and non-transferable process acquired through reflection and a knowledge being collected, deepened, contemplated, recorded, and disseminated in time upon a certain social consensus (Barité, 2001; Dahlberg, 1993).

In a previous study (Guimarães, Sales, de, Martínez-Ávila, & Alencar, 2014), working with ISKO conference proceedings, it was possible to observe that KO has been mainly understood as a theoretical-practical field whose object is knowledge, especially recorded knowledge, as well as the concepts and their structures, that are instrumentally stated by the knowledge organization systems (KOS). Its most relevant processes are related to classification and indexing, as well as to information retrieval. Its methodology is deeply connected to the systematization and modeling of concepts of recorded scientific knowledge as discursive formations and sociocultural practices. Classification systems, thesauri, controlled vocabularies, ontologies, and citation indexes are considered important KO tools.

If, on the one hand, it is possible to observe a certain level of synergy and complementarity in conceptual terms in the KO scientific community, on the other hand, it is verified a high dynamicity of the field, revealing a rich ensemble of theoretical influences along the time, from an ontological and Aristotelian approach to a cultural approach that considers different and dynamic slants.

In this sense, this chapter draws from the assumption that slants are inherent to any organization of knowledge and that they are manifested in multiple dimensions. In this chapter, the term slant is used in accordance to the Oxford English Dictionary as "a particular point of view from which something is seen or presented" (Slant, 2017).

The purpose of this chapter is to characterize KO as a naturally slanted field that is affected by geographic and diachronic variations in such a way that the recognition of a *slanted KO* could be considered an ethical option in the KO theory and practice. This recognition is a necessary consequence for the coexistence of diverse groups and communities, with local characteristics, meanings, and idiosyncrasies, that will need to communicate with each other in global information systems of information.

6.2. KO as a Dynamic Field

A diachronic approach to the KO field, especially based on the ISKO history, shows a turning point that has been occurring during the last two decades, from an ontological perspective to a more epistemological one (Arboit, 2014). This situation evidences what Mai (2013, p. 251) considers a transition "from an essentialistic conception of a universe of knowledge to a pluralistic, pragmatic view of domains and practice as its foundation."

Historically, publications in the KO field, especially in the ISKO environment, were built in an ontological perspective of KO, based on the nature of being, becoming, existence, or reality, as well as the basic categories of knowledge and their relations led to the approach to KO as a set of concepts. In this sense, Ranganathan's classification theory and Dahlberg's theory of concept can be considered foundational representatives of this conception (Dahlberg, 1978a, 1978b, 1993, 1994, 1995). An important example of this approach is Dalhberg's Classification Scheme for Knowledge Organization Literature, described by herself as a bibliographic control tool for the KO field, whose structural sequence is represented by a matrix 3×3 of main classes, called "Systematifier" (Dahlberg, 1993). After more than one decade, the KO scholars started to be more concerned with the epistemological basis of this field, as a way to better understand its dynamics.

6.2.1. The Epistemological Perspective in KO

The epistemological perspective of KO is more concerned on the nature, steps, and limits of human knowledge, especially within the relations established between the subject and the object of knowledge. KO would be a dynamic environment. In this transition, it is also possible to observe a conceptual migration of knowledge, from a "subjectively and objectively conclusive certainty of the existence of a fact or a state of a case" that is not transferrable and can only be acquired through a personal rethinking (Dahlberg, 1995, p. 10), to something that is "a social product, a social need and a social dynamo" (Barité, 2001, p. 42).

This epistemological perspective, led by Hjørland, focuses on a more sociocognitive approach and incorporates issues related to the user and KO contexts (Hjørland & Albrechtsen, 1995; Hjørland & Nicolaisen, 2010; Hjørland, 2002a,b, 2003, 2009, 2010, 2013). To achieve this, Hjørland proposes a new interpretation for the concept theory in which he states that "the best understanding and classification of theories of concepts is to view and classify them in accordance with epistemological theories (empiricism, rationalism, historicism, and pragmatism)" (Hjørland, 2009, p. 1519). The author, referring to what Margolis and Laurence regard as the core readings in concept theory, presents five perspectives that have to be taken into account when discussing the concept theory: the history of concepts, the hermeneutics, the cultural historical activity theory, the pragmatism, and the speech-act theory. In addition, Hjørland (2009) summarizes the major theories of concepts as they are presented and discussed in the contemporary literature: the classical (or Aristotelianism) theory of concepts, the probabilistic one, the theory-based theory of concepts, the neoclassical theory of concepts, and the conceptual atomism. The author concludes that the historicist and pragmatist understandings of concepts are the most fruitful views as part of a broader paradigm in Information Science, and also considers concepts as clearly social constructs based on socially negotiated meanings "that should be identified by studying discourses rather than by studying individual users or a priori principles" (Hjørland, 2009, p. 1530).

More specifically, in this sociocognitive approach, two complimentary and dialogical perspectives can be highlighted: the domain analysis and the semiotic perspective. These perspectives are not mutually exclusive, but rather complementary.

The perspective of domain analysis is directly linked to context and discursive communities in which KO operates, in such a way that those discursive communities emerge from the social division of labor, and reflect, through their members, a "labor ecology." As a consequence, the concept of KO, under a domain analysis perspective, is a product of interpretation and can only be understood in a given context and by a given community, hence that the same knowledge can be understood differently by other communities thus reflecting pluralism (Hjørland, 2002a; Hjørland & Albrechtsen, 1995; Smiraglia, 2011, 2012; Tennis, 2003).

About the semiotic perspective, Jens-Erik Mai, in his seminal paper, begins with the central problem of understanding how the interpretation of a document's subject matter can be represented to make it possible for others to gain access to the document. In this sense, the author demonstrates how KO representation processes characterize a set of multiple interpretations, and evidence a typical Peircean unlimited semiosis, "which could be seen as the connecting of sign or the process of one sign producing another sign" (Mai, 2001, p. 598). The semiotic perspective is concerned with reception issues and significance in the KO context (Mai, 2000, 2004, 2009).

Considering information as a possibility to give "means" for "knowing," as well as considering that the main goal of KO is to enable users to retrieve recorded knowledge in order to satisfy their information needs, it is essential to recognize the existence of a communicational process that is constantly modified and reinterpreted according to different situations (Thellefsen, Thellefsen, & Sørensen, 2013). In this sense, and according to Peircean semiotics, the knowledge representation process has its own dynamics, "which must suggest the creation of interpretants that are mutually related in a continuous process of signification" (Almeida, Fujita, & Reis, 2013, p. 239).

The recognition of a plurality of users and information needs as well as plurality of interpretations brings space to the recognition of culture as a determinant aspect of KO.

6.2.2. *The Cultural Perspective in KO*

The cultural perspective in KO can be considered as a derivation of the epistemological perspective. In that perspective, historicity, time and space conditions and idiosyncrasies of the communities involved in the production, and use of knowledge play a leading role. Under this perspective, KO processes, tools, and products suffer a direct impact from cultural diversity and present an inherent ethical dimension.

In this field, perhaps the first work that points out the need for a cultural approach in knowledge organization and representation is the book *Prejudices and Antipathies* (Berman, 1971), where the author analyzes dozens of Library of Congress subject headings (in topics such as race, religion, and sex) in order to evidence their unsuitability to different cultural standards in a growing plural society. In his work, the author highlights the inherent cultural nature of KO, in which it is impossible to adopt "neutral" representations because this supposed impartiality can lead to discriminations, lack of visibility, or even cultural domination in different cultures.

Antonio García Gutiérrez, concerned on the effects of cultural diversity in KO, proposes what he calls an "interactive epistemography" with the inclusion of different cultural views and their relationships, establishing a transcultural ethics of mediation (García Gutiérrez, 2002).

Michèle Hudon observes the need for equal treatment to the languages involved in multilingual thesauri, which is not limited to the identification of a similar concept, but includes thinking of possible solutions to administrative, linguistics, and semantics problems, as well as problems of technological nature (Hudon, 1997) because languages are more than just a set of words and rules put together; they reflect the cultural values of their speakers (Hudon, 1997, 1999).

Clare Beghtol (2002, 2005) proposes the theoretical concepts of cultural warrant and cultural hospitability to ethically support a KOS, in such a way that this system should reflect the users' values and to be opened to cultural changes along the time. In this sense, the KOS will be able to ideally accommodate the various warrants of different cultures and to appropriately reflect the assumptions of any individual, group, or community.

Maria José López-Huertas, while referring to some current research questions in KO, points out the multilingual, cross-cultural, and social issues in KO as an important research topic and lists some relevant — and urgent — cultural challenges: How to manage multilingual and sociocultural diversity? How to strive toward universality without sacrificing diversity? How to integrate sociocultural differences in KO? How to turn cross-cultural issues into a desired quality in KO? How minorities and marginalized sectors are taken into account in general systems? (López Huertas, 2008).

Grant Campbell, Fábio Pinho, and José Augusto Guimarães have discussed the cultural damages caused by the "tabula rasa" of universal KOS to specific domains, for example, the male homosexuality. In this sense, Campbell (2000) discusses the tensions that can occur between aboutness and meanings in such a way that aboutness cannot be considered a stable document content anymore, but dependent on the system designer's social and ideological perspectives. Thus, categories become fluid and unstable due to the fact that community members want to see themselves as socially constructed, and at the same time they also want to be a special category. Those problems directly affect and reconceptualize precision in a way that it can be no longer considered a mere information retrieval measure, but the capability of "making sense," that is, of including different cultures. For that, the recognition of metaphors as vehicles of specific meanings for specific discursive communities is an important variable that has to be considered in KOS (Pinho & Guimarães, 2012).

Therefore, in this cultural perspective, the issue of ethics decisively emerges and knowledge organization and representation takes the role of assigning surrogates to informative content, coming from a "power to name" (Olson, 2002) that has been granted by society to the information professional. The power to name evokes the situational and political stance of KOS and emphasizes the figure of heterogeneous users in search for voice in library catalogs (Guimarães, Pinho, & Milani, 2016). From this perspective, in which Hope Olson plays an effective central role, the following characteristic elements can be drawn up: (a) the recognition of diversity as something inherent to KOS; (b) the recognition of the impact of human and social factors into the KO activities; (c) the conception of KO domain as a social product; (d) the suggestive (and not prescriptive) nature of KOS; (e) the KOS as tools to promote global dialogue; (f) the historical migration from Universal systems to Global systems; and (g) the movement from standardization to the promotion of intercommunication and, as a consequence, from vocabulary control to interoperability.

This scenario leads to the recognition that every knowledge organization or representation is committed to a certain world view and, as a consequence, is slanted. The recognition of those different worldviews presupposes the recognition of different approaches, according to the different cultures involved, in time and space. Those approaches have to be considered in the description processes because they will be also present in the search processes. These searches are many times affected and influenced by day-by-day situations and the social-linguistic characteristics of each community. In this sense, these aspects must be considered as an ethical commitment by KO, by a *slanted KO*.

6.3. KO as a *Slanted* Field

Although only in recent decades KO has been discussing the impossibility of neutral, unidirectional, and standardized representations due to the existence of inherent slants, Ranganathan recognized them in 1964, when he created the journal *Library Science with a Slant to Documentation*.

Lancaster (1991) referred to the abstracting process as something that is necessarily slanted to the users' needs. In this sense, the term slant incorporates the idea of commitment, by pursuing a certain goal for clear purposes and, thus, it cannot be considered something negative but as something useful and desirable.

A slant can be considered a particular point of view from which something is seen or presented as well as the search for another direction — most of the times diagonal or transversal — than the traditional Cartesian ones. In a positive context, slanting or tendency (Hjørland, 2008; McIlwaine, 2003) conveyed by the document surrogates to ensure specificity to particular discursive communities or knowledge domains can be observed.

The inherent existence of slants in KO processes, tools, and products derives from the fact that if KO, under a cultural paradigm, presupposes the socialization of knowledge, it is mandatory to recognize that this socialization process is embedded by different values or cultural standards that will be transformed into slants. In other words, every KOS is naturally committed to a certain world view and the assumption of a set of values and beliefs.

The problem occurs when those slants act as bias, by means of an intentional or unintentional "detour" (Milani, 2014; Milani, De, Guimarães, & Olson, 2014).

Unlike slants, biases often assume a negative connotation — a distortion — due to the fact that they traditionally occur through beliefs or attitudes that justify — or evidence — any kind of unfairness that can cause dangerous consequences such as segregation or harm.

Guimarães (2006) stresses two special kinds of dangerous biases: prejudice and proselytism. Prejudice can occur, for instance, when a KOS establishes a dichotomic categorization between Christian and non-Christian religions, as if the Christians were the rule and the non-Christians were exceptions or odd options. Another situation occurs, for example, when an information professional, considering the inexistence of specific descriptors for the topic gay night clubs, decides to insert the book under the descriptor promiscuous environments. Prejudice presupposes the inexistence of adequate or prior knowledge about something leading to an unreasonable view based on preconceived ideas.

Prejudice can also occur through elaborated metaphors, creating euphemisms, true "lupi in vestimentum ovium," when supposedly gentle and neutral words are used to hide a negative judgment. This is the case of expressions such as "chocolate" for black men, or even "very well-educated man" for gay man.

Proselytism derives from the attempt to convert people to a different point of view that can be religious, political, ideological, etc. The mentioned example of Christian and non-Christian religions can also evidence a proselytism of Christianism, or even the use of the term "profanes" to designate non-freemason men. Another example is the use of the generic term Evangelical to designate all the religions derived from Protestantism, because this generic term comprises not only solid doctrinal-based churches, but also other churches that have been created as a consequence of internal fights and, sometimes, contradict some basic Luther's principles such as the prohibition of selling indulgences.

Slants in KO occur in different dimensions: Slants by the author/document, Slants of the indexer/classifier, Slants of the KOS/classifications, Slants of the environment (e.g., institution), and Slants of the user. And they are supposed to coexist in relative harmony even with some tensions or oppositions that can occur among them, because every instance needs to have a voice in this process.

Every KO procedure, tool, or product is slanted because it is located in a certain space, in a certain time and arises from a set of ideological positions. So, prejudices in a certain moment might have been moral values in another period, and virtues in a given social context might be translated into sins in another context. Those questions suggest the importance of time and space as important contextualizing factors — or axes — in a slanted KO.

6.4. Time and Space as Axes for a Slanted KO

In 2003, Joseph Tennis wrote a paper entitled "Two Axes for Domain Analysis," where he proposed two analytical devices — axes — that could be used to define and to delineate a domain as well as to operationalize domain analysis. Through those axes, it was possible to conceive a domain as a kind of Cartesian coordinate plane on which elements occupy a certain position and interact with each other (Tennis, 2003, 2016).

Based on this approach, it is also possible to think about another Cartesian coordinate plane that contextualizes knowledge organization and representation in order to avoid the occurrence of naive conceptions, or even prejudices and proselytisms derived from the situations previously pointed out by Guimarães (2006) as dicotomic categorizations, narrow worldviews, terminological haziness, decontextualized polysemic representations, and the artificiality of neologisms derived from political correctness.

For this purpose, we propose the time and space approach as more than mere facets of subject representation, but mainly as contextualizing facets that can contribute to establish bridges between local and global needs.

In order to give examples of this situation, we divide the discussion into two dimensions: the conceptual dimension, more related to KO processes, and the terminological dimension, more related to knowledge representation process.

6.4.1. Conceptual Dimension of KO Slants

The conceptual dimension of time and space slants in KO occurs when the concept (attributes, characteristics, nature, etc.) varies in time and space, although the representational terminology remains the same.

An example, referring to a time slant, is the term *Marriage* that, in the 19th Century, was related to a civil or religion union between a man and a woman but, in this 21st century, is related to a variety of civil or religion unions between a man and a woman, a man and a man, a woman and a woman, etc. In this sense, Tennis (2012a, 2012b) discusses the issue of subject ontogeny (subject across time) with the case of eugenics in the DDC all over the time. To achieve this, the mentioned author refers to Michael Buckland's concept of meaning (Buckland, 2012, p. 158) for whom "meanings are established by usage, and so always draw on the past." The librarian, then, is creating descriptions by drawing on the past, but expressing them with an eye to the future. In another paper, Tennis (2016) remains concerned on how to preserve the functionality of classification schemes over time (p. 574) in such a way that "schemes built for a particular purpose can be deployed in service to another context" (p. 574). To achieve this, the author presents ontological and epistemic concerns that reaffirm the need of a diachronic approach to the analysis of KO systems: the negotiated ontological commitments in such a way that "we only know the subject through context" (p. 576) and the epistemic problem of "knowing the subject" in such a way that "we can know the presence of subject by their contexts" (p. 577).

Based on daily situations, frequently experienced by people from different cultures immersed in a global and inter-connected communication scenario, some examples can be presented as a part of their local day-by-day experiences and values, that can be reflected in users' searches in information systems.

The conceptual space (here understood in a broader sense, as context) occurs, for example, with the term *Abortion* which, in the Roman Catholic Church and Catholic countries, for instance, can be conceived and understood as a sin and as a kind of homicide, whereas according to the

Women's Rights Movements, it is considered a right of the woman over her own body. Another example comes from the concept of *Life after death* in different religious settings and communities, in which it achieved through many Supreme beings, for Polytheists, and through one only Supreme being for Monotheists (Yahveh, Adonai, or Jehovah for the Jews, according to different periods, Allah for Muslims, Christ for Christians). And among the Christians, the death after life can be obtained through Christ by faith and good works (for the catholics) and only by grace and faith (for the protestants). These concepts could affect, for instance, when someone searches books on Salvation.

The conceptual time slant occurs when a term assumes different meanings or even perceptions along the time. An example for this is *Polygamy* among Mormons: it was accepted in a certain way in the 19th century, but currently is not accepted anymore. It is interesting that this same term — *Polygamy* — also presents a space slant because it can be considered a crime in many Western societies while it can be considered a social standard in many Eastern societies.

6.4.2. Terminological Dimension of KO Slants

The terminological dimension of time and space slants in KO occurs when the concept (attributes, characteristics, nature, etc.) possesses a certain stability in time and space, although the representational terminology changes as time goes by, in most cases due to the phenomenon of the "politically correctness." For example, *Gay couples* → *Same sex couples* → *LGBT parenting*. In this case, it happens a space terminological slant situation because while in certain countries, the term *Gay couples* is considered usual and respectful; in other countries it can be considered derogatory and the most suitable term is *Homosexual couples* or even *Homoaffective couples*.

Considering this scenario, some additional concerns arise about the possible lack of cultural warrant related to the effects of the politically correctness in KO and the metaphorical (and metonymic) dimension of KO. In the first case, we need to consider the importance of political correctness as a factor to promote social inclusion and to avoid segregations. However, it is also important to consider that this phenomenon needs to be socially built in a way to have the desired representativeness in different cultures, otherwise those terms would be completely artificial. A good example is the term *Esthetically challenged*, used to represent ugly persons. It maintains the same problem of the term ugly since the esthetic perception is not homogeneous. In the same line are the terms *Persons of size* (for fat persons) or *Hymenally challenged* (for nonvirgin women).

In the second case, and in direct relation with the political correctness, some figures of speech are used to represent concepts (most of the times through metaphors or metonyms) and assume different conceptual perceptions in different contexts (and, in some of them, they assume an inaccurate or a prejudiced meaning). An example is the use of the term African-Americans to designate Black persons because it immediately raises a question: What about the Dutch descendants in South Africa? And the Portuguese descendants in Angola? And the Arab descendants in North Africa? This term ignores the fact that Africa is a continent formed by different ethnic groups with different skin colors.

Another example occurs with the use of the term Hispanics to designate Latin-Americans, as it is usual in the United States administration. In this sense, some questions arise: Are Brazilians Hispanic? Aren't Spaniards Hispanic? This example ignores the fact that Spain colonized a great part of South and Central America (but not the whole) and some other colonizer countries also existed in the same regions as Portugal, Netherlands, France, and England.

A third example would be found in the use of the term America to designate the United States (e.g., President Donald Trump's slogan: "Make America great again"). This term ignores the fact that America is a huge continent geographically composed of three parts (North, Central, and South America) and ethnically composed of two major groups (Latin America and Anglo-Saxon America). In this case, the use of the term America to designate only the United States is an appropriate metonym that segregates a huge group of 34 countries, from Canada to Argentine. Is America a single country or a huge continent?

6.5. Conclusion

As we can see, going back to the objectives of this chapter, KO can be considered a naturally slanted field that is affected by contextual variations (author's context, user's context, indexer's context, KOS' context, etc.), that present both geographic and diachronic natures. In this sense, time and space are important axes to be considered in KO procedures, tools, and products, in order to contextualize the information content (by recognizing local realities) and to avoid the conception of "universal" categorizations as the only alternative. The recognition of time and space variables of a term and/or a concept in KOS, for example, as related terms (or, even in a deeper movement, in contextual scope notes) can be a useful tool for an intercultural dialogue in a global world.

Considering these aspects, the recognition of a *slanted KO* could be considered an ethical option in the KO theory and practice, due to the fact that the intervention of an ethical analysis in KOS goes further of being appropriate, to be especially necessary, as it tries to intervene in represented and possibly disseminated biases practiced so far, in addition to improving representation tools so that an awareness of how and why they are made is risen. This recognition is a necessary consequence for the coexistence of diverse groups and communities, with local characteristics, meanings, and idiosyncrasies, that will need to communicate with each other in global information systems of information.

In summary, the conception of a slanted KO as a starting point for the development of KO processes, tools, and products is a way to consider the power to name as a necessary professional concern (Olson, 2002), and to promote an intercultural ethics of mediation (García-Gutiérrez, 2002), that is culturally warranted (Beghtol, 2002, 2005), in order to avoid cultural damages (Campbell, 2000; Pinho & Guimarães, 2012). This slanted KO is also a valid way to guarantee that descriptions can reflect the past while keeping "an eye to the future" (Buckland, 2012), based on classification schemes whose functionality remains over time (Tennis, 2016).

Acknowledgments

I would like to thank Daniel Martínez Ávila, Suellen Milani, Natalia Tognoli, and Isadora Victorino Evangelista (São Paulo State University) for her valuable comments and suggestions.

References

Almeida, C. C., Fujita, M. S. L., & Reis, D. M. (2013). Peircean semiotics and subject indexing: Contributions of speculative grammar and pure logic. *Knowledge Organization*, *40*(4), 225–241.

Andersen, J. (2002). Communication technologies and the concept of knowledge organization. *Knowledge Organization*, *29*(1), 29–39.

Arboit, A. E. (2014). O processo de institucionalização sociocognitiva do domínio de organização do conhecimento a partir dos trabalhos científicos dos congressos da ISKO. Dissertation of PhD in Information Science. São Paulo State University – UNESP, Marília.

Barité, M. (2001). Organización del conocimiento: un nuevo marco teórico-conceptual en Bibliotecología y Documentación. In K. Carrara (org.) *Educação, Universidade e Pesquisa* (pp. 35–60). Marília: Unesp-Marília-Publicações.

Beghtol, C. (2002). A proposed ethical warrant for global knowledge representation and organization systems. *Journal of Documentation*, *58*(5), 507–532.

Beghtol, C. (2005). Ethical decision-making for knowledge representation and organization systems for global use. *Journal of the American Society for Information Science and Technology*, *56*(9), 903–912.

Berman, S. (1971). *Prejudices and antipathies: A tract of Library of Congress Subjects Headings concerning people*. Metuchen, NJ: Scarecrow Press.

Buckland, M. (2012). Obsolescence in subject description. *Journal of Documentation*, *68*(2), 154–161.

Café, L., & Brascher, M. (2011). Organização do conhecimento: teorias semânticas base para estudo e representação de conceitos. *Informação e Informação*, *16*(3), 25–51.

Campbell, G. (2000). Queer theory and the creation of contextualized subject access tools for gay and lesbian communities. *Knowledge Organization*, *27*(3), 122–131.

Dahlberg, I. (1978a). A referent-oriented, analytical concept theory for INTERCONCEPT. *International Classification*, *5*(3), 142–151.

Dahlberg, I. (1978b). Teoria do conceito. *Ciência da Informação*, *7*(2), 101–107.

Dahlberg, I. (1993). Knowledge organization: Its scope and possibilities. *Knowledge Organization*, *20*(4), 211–222.

Dahlberg, I. (1994). Conceptual structures and systematization. In G. Negrini (Ed.), *Categorie, Ogetti e Strutture della Conoscenza*. Roma: Consiglio Nazionale dele Ricerche. Istituto Nazionale sulla Ricerca e Documentazione.

Dahlberg, I. (1995). Current trends in knowledge organization. In F. J. García Marco (Ed.), *Organización del conocimiento em sistemas de información y documentación* (pp. 7–26). Zaragoza: Universidad de Zaragoza.

Dahlberg, I. (2006). Knowledge organization: A new science? *Knowledge Organization*, *33*(1), 11–19.

Furner, J. (2009). Interrogating "identity": A philosophical approach to an enduring issue in knowledge organization. *Knowledge organization*, *36*(1), 3–16.

García Gutiérrez, A. (2002). Knowledge organization from a 'Culture of the Border' towards a transcultural ethics of mediation. In *Proceedings of the seventh international ISKO conference*, July, Granada, Spain, p. 7.

García Marco, F. J. (1997). Advances en organización del conocimiento en España: los II encuentros sobre organización del conocimiento en los sistemas de información y documentación. In *Actas del II Encuentro de Isko-España*, November, Zaragoza, Spain, p. 2.

Gilchrist, A. (2011). A note on the theory and practice of knowledge organization and knowledge representation. *Scire*, *17*(1), 23–28.

Gnoli, C. (2008). Ten long-term research questions in knowledge organization. *Knowledge organization*, *35*(2/3), 137–149.

Gnoli, C. (2011). Ontological foundations in knowledge organization. *Scire*, *17*(1), 19–34.

Guimarães, J. A. C. (2006). Aspectos éticos em organização e representação do conhecimento (ORC): uma reflexão preliminar. In M. N. González Gómez & E. G. D. Orrico (Eds.), *Políticas de memória e informação: reflexos na organização do conhecimento* (pp. 237–264). Natal: EDUFRN.

Guimarães, J. A. C., Pinho, F. A., & Milani, S. O. (2016). Theoretical dialogs bout ethical issues in knowledge organization: García Gutiérrez, Hudon, Beghtol, and Olson. *Knowledge Organization*, *43*(5), 338–350.

Guimarães, J. A. C., Sales, R., de, Martínez-Ávila, D., & Alencar, M. F. (2014). The Conceptual Dimension of Knowledge Organization in the ISKO Proceedings Domain: A Bardian Content Analysis. In *Proceedings of the Thirteenth International ISKO Conference*, May, Würzburg, Germany.

Hjørland, B. (2002a). Epistemology and the socio-cognitive perspective in information science. *Journal of the American Society for Information Science and Technology*, *53*(4), 257–270.

Hjørland, B. (2002b). Meta-analysis should also be visible inside information science (letter to the editor). *Journal of the American Society for Information Science and Technology*, *53*(4), 324.

Hjørland, B. (2003). Arguments for epistemology in information science. *Journal of the American Society for Information Science and Technology*, *54*(8), 805–806.

Hjørland, B. (2008). Deliberate bias in knowledge organization? In Proceedings of the tenth international ISKO conference, Germany, Würzburg, p. 10.

Hjørland, B. (2009). Concept theory. *Journal of the American Society for Information Science and Technology*, *60*(8), 1519–1536.

Hjørland, B. (2010). The foundation of the concept of relevance. *Journal of the American Society for Information Science and Technology*, *61*(2), 217–237.

Hjørland, B. (2013). Citation analysis: A social and dynamic approach to knowledge organization. *Information Processing and Management*, *49*(6),1313–1325.

Hjørland, B., & Albrechtsen, H. (1995). Toward a new horizon in Information Science: Domain-analysis. *Journal of the American Society for Information Science and Technology*, *46*(6), 400–425.

Hjørland, B., & Nicolaisen, J. (2010). The social psychology of information use: Seeking "friends", avoiding "enemies". *Information Research*, *15*(3). Retrieved from http://www.informationr.net/ir/15-3/colis7/colis706.html

Hudon, M. (1997). Multilingual thesaurus construction: Integrating the views of different cultures in one gateway to knowledge and concepts. *Knowledge Organization*, *24*(2), 84–91.

Hudon, M. (1999). Accessing documents and information in a world without frontiers. *The Indexer*, *21*, 156–159.

Jaenecke, P. (1994). To what end knowledge organization? *Knowledge Organization*, *21*(1), 3–11.

Kiel, E. (1994). Knowledge organization needs epistemological openness. *Knowledge Organization*, *21*(3), 148–152.

Lancaster, F. W. (1991). *Indexing and abstracting in theory and practice*. London: Library Association.

López-Huertas, M. J. (2008). Some current research questions in the field of knowledge organization. *Knowledge Organization*, *35*(2–3), 113–136.

Mai, J.-E. (2000). *The subject indexing process: An investigation of problems in knowledge representation (PhD in Information Science)*. Austin, TX: University of Texas at Austin.

Mai, J.-E. (2001). Semiotics and indexing: An analysis of the subject indexing process. *Journal of Documentation*, *57*(5), 591–622.

Mai, J.-E. (2004). Classification in context: Relativity, reality, and representation. *Knowledge Organization*, *31*(1), 39–48.

Mai, J.-E. (2009). Socio-cognitivism (domain analysis). In *ASIST annual meeting*, Vancouver, Canada.

Mai, J.-E. (2013). Ethics, values and morality in contemporary library classifications. *Knowledge Organization*, *40*(4), 242–253.

McIlwaine, I. C. (2003). Trends in knowledge organization research. *Knowledge Organization*, Würzburg, *30*(2), 75–86.

Milani, S. De, O., Guimarães, J. A. C., & Olson, H. A. (2014). Bias in subject representation: Convergences and divergences in the international literature. In W. Babik (org.), *Knowledge organization in the 21st century: Between historical patterns and future perspectives* (pp. 335–344). Wurzburg: Ergon Verlag.

Milani, S. O. (2014). *Bias na representação de assunto: uma discussão de oposições binárias nos Functional Requirements for Subject Authority Data (FRSAD) (Dissertation of PhD in Information Science)*. São Paulo State University – UNESP, Marília.

Olson, H. A. (2002). *The power to name: Locating the limits of subject representation in libraries*. Dordrecht: Kluwer Academic.

Pinho, F. A., & Guimarães, J. A. C. (2012). Male homosexuality in Brazilian indexing languages: Some ethical questions. *Knowledge Organization*, *39*(5), 363–369.

Slant (2017). *Oxford English Dictionary*. Oxford: Oxford University Press. Retrieved from https://en.oxforddictionaries.com/definition/slant

Smiraglia, R. P. (2011). Domain coherence within knowledge organization: People, interacting theoretically, across geopolitical and cultural boundaries. In *Proceedings of Annual CAIS/ACSI conference*, Fredericton, Canada, p. 39.

Smiraglia, R. P. (2012). Epistemology of domain analysis. In H.-L. Lee & R. P. Smiraglia (Eds.), *Cultural frames of knowledge* (pp. 111–124). Wurzburg: Ergon.

Tennis, J. T. (2003). Two axes of domain analysis. *Knowledge Organization*, *30*(3–4), 191–195.

Tennis, J. T. (2008). Epistemology, theory, and methodology in knowledge organization. *Knowledge organization*, *35*(2–3), 102–112.

Tennis, J. T. (2012a). Subject Ontogeny: Subject Access through Time and the Dimensionality of Classification. *Proceedings of the Seventh International ISKO Conference*, Würzburg: Ergon, 8.

Tennis, J. T. (2012b). The strange case of eugenics: a subject's ontogeny in a long-lived classification scheme and the question of collocative integrity. *Journal of the American Society for Information Science and Technology*, *63*(7), 1350–1359.

Tennis, J. T. (2016). Methodological challenges in scheme versioning and subject ontogeny research. *Knowledge Organization*, *43*(8), 573–580.

Thellefsen, M., Thellefsen, T., & Sørensen, B. (2013). A pragmatic semeiotic perspective on the concept of information need and its relevance for knowledge organization. *Knowledge Organization*, *40*(4), 213–224.

Zins, C. (2004). Knowledge organization: An epistemological perspective. *Knowledge Organization*, *31*(1), 49–54.

About the Editors

Jack Andersen is Associate Professor, PhD, at the Royal School of Library and Information Science, University of Copenhagen. Andersen's research interests are centered on digital media, classification, and genre theory. Andersen has published several articles dealing with how to understand knowledge organization, scholarly literature, and information literacy from social, theoretical, epistemological, and genre theoretical approaches. His recent edited volume, *Genre Theory in Information Studies*, explores different dimensions of genre in analyses of information. He is on the advisory board for the international research network, *Genre across Borders*. In 2009–2013, Andersen served as vice-director and chair of Department at the Royal School of Library and Information Science. He teaches courses on the theory of science, academic writing, media theory, and digital culture.

Laura Skouvig is Associate Professor, PhD, at the Royal School of Library and Information Science, University of Copenhagen. Her primary research interest is in the field of information history focusing on the history of the information age and information society. She has published articles on information networks, surveillance and communication of information primarily in the period of late 18th and early 19th century. Theoretically, she works with inspiration from Michel Foucault and rhetorical genre theory. She is on the editorial boards of two international journals: *Library and Information History* and *Information and Culture*. She teaches courses on the theory of science and knowledge culture.

Index

Printed and bound by CPI Group (UK) Ltd, Croydon, CR0 4YY
13/09/2026
14961058-0006